Physics in Your Life

Richard Wolfson, Ph.D.

THE
GREAT
COURSES®

PUBLISHED BY:

THE GREAT COURSES
Corporate Headquarters
4840 Westfields Boulevard, Suite 500
Chantilly, Virginia 20151-2299
Phone: 1-800-832-2412
Fax: 703-378-3819
www.thegreatcourses.com

Richard Wolfson, Ph.D.

Professor of Physics, Middlebury College

Richard Wolfson is Benjamin F. Wissler Professor of Physics at Middlebury College, where he has also held the George Adams Ellis Chair in the Liberal Arts. He did undergraduate work at MIT and Swarthmore College, graduating from Swarthmore with a double major in physics and philosophy. He holds a master's degree in Environmental Studies from the University of Michigan and a Ph.D. in Physics from Dartmouth. Professor Wolfson's published work encompasses such diverse fields as medical physics, plasma physics, solar energy engineering, electronic circuit design, observational astronomy, theoretical astrophysics, nuclear issues, and climate change. His current research involves the sometimes violently eruptive behavior of the Sun's outer atmosphere, or corona. He also continues an interest in environmental science, especially global climate change.

Dr. Wolfson is particularly concerned with making science relevant to nonscientists and to students from all walks of academic life. His textbook, *Physics for Scientists and Engineers* (Addison Wesley, 1999), is now in its third edition and has been translated into several languages. His books *Nuclear Choices: A Citizen's Guide to Nuclear Technology* (MIT Press, 1993) and *Simply Einstein: Relativity Demystified* (W.W. Norton, 2003) exemplify Wolfson's interest in making science accessible to nonscientists. He has also published in *Scientific American* and has produced videotaped courses for The Teaching Company, including *Einstein's Relativity and the Quantum Revolution: Modern Physics for Nonscientists*, *Energy and Climate: Science for Citizens in the Age of Global Warming*, and *Physics in Your Life*. Professor Wolfson has spent sabbaticals at the National Center for Atmospheric Research in Boulder, Colorado; at St. Andrews University in Scotland; and at Stanford University.

Table of Contents
Physics in Your Life

Physics in Your Life

Scope:

Physics is the science that governs the workings of physical reality at its most fundamental level. Thus, physics is important in understanding the ultimate nature of the Universe. But it is equally important in our everyday lives. The commonest actions—such as walking, breathing, or driving a car—are all based on principles of physics. The natural world delights us with a host of physics-based phenomena, from rainbows and snowflakes to the blue of the sky, the daily rotation of our planet, and the celestial companionship of the orbiting Moon. And physics-based technology is ubiquitous in modern life—from the CDs and DVDs that entertain and inform us to the antilock brakes that make driving safer, the global positioning system that helps us navigate about our planet, medical imaging that enhances our health, microwaves that cook our food, airplanes that transport us swiftly about the globe, lasers that scan our supermarket purchases, and the semiconductor electronics at the heart of our computers, cell phones, digital cameras, personal digital assistants, and audiovisual systems.

This course introduces principles of physics through their application to everyday life. It's more than a course in physics and more than a laundry list of "how things work." Rather, it combines the two, offering a back-and-forth interplay between everyday applications of physics and the physics principles needed to understand them. Applications include the simplest everyday activities, natural phenomena that affect our everyday lives, and especially, modern technology. Physics principles covered range from Newton's laws of motion, known for hundreds of years but still vitally relevant, to concepts from atomic and quantum physics that underlie such diverse technologies as semiconductor electronics, lasers, and medical imaging.

The course is organized into six modules of six lectures each. The first five modules deal with specific realms of physics and related applications; the sixth is a potpourri of physics applications that draw from more than one of the earlier modules. Although there's no obviously straightforward path through the myriad applications included here, the lectures are designed to build on each other and to flow smoothly from one to the other. *Physics in Your Life* is aimed at intelligent nonscientists, and the presentation of physics concepts and applications is entirely nonmathematical.

Given that the course is presented on audiovisual media, the first module—**"Sight and Sound"**—begins with the technology behind CDs and DVDs, not only explaining how these work but raising questions that lead to the basic principles of light and sound. Subsequent lectures explore these principles in application to such diverse topics as rainbows, optical fibers for communications, musical instruments, and laser vision correction.

"Going Places" looks at motion and its applications, from simple walking to modern automobile technology, airplane flight, and space travel. This module is based on Newton's laws, generalized to include such diverse topics as fluid motion, conservation of energy, and the dynamics of space flight from communications satellites to interplanetary exploration.

"Plug In, Turn On" explores the electromagnetic basis of so many contemporary technologies and natural phenomena. Matter itself sticks together through electrical interactions, while the intimate relations between electricity and magnetism are at the heart of technologies ranging from electric motors and generators to videotapes and credit cards. We've learned numerous ways to produce electrical energy, and we respect not only the good it can do but also the dangers it poses. Finally, electricity and magnetism join to make possible the electromagnetic waves that include visible light and that enable the growing host of wireless technologies in our everyday lives.

"From Atom to Computer" starts with basic ideas in atomic physics and builds through explanation of the semiconductor materials at the basis of today's electronics; the operation of the transistors that are the essential electronic components; the agglomeration of close to a billion transistors in today's most sophisticated integrated circuits; the simple logical operations that underlie all our computers can do; and the structural organization of an entire computer.

The fifth module, **"Fire and Ice,"** introduces concepts and applications related to heat. Lectures range from "Physics in the Kitchen" to "Life in the Greenhouse," the latter a look at how principles of thermal physics establish Earth's climate and how we humans may be changing that climate. "The Tip of the Iceberg" describes the thermal response of materials, including the unusual behavior of water in both liquid and solid form—a behavior with enormous implications for life on Earth. The module ends with a look at humanity's appetite for energy in the face of limitations posed by the laws of thermodynamics. Those laws themselves are the subject of "Like a

Work of Shakespeare," a lecture so titled because one such law has been deemed as essential to an educated person as the Bard's writings.

The sixth module is a potpourri of applications and principles from the specific to the cosmic. "Your Place on Earth" describes the workings and increasingly pervasive applications of the space-based Global Positioning System (GPS). "Dance and Spin" looks at applications of rotational motion from dance to pulsars. "The Light Fantastic" explores the operation of lasers and their applications in everything from supermarket scanners to the Internet. "Nuclear Matters" looks at the behind-the-scenes roles that nuclear physics plays in our lives. "Physics in Your Body" emphasizes modern medical techniques, such as MRI and particle-beam treatments for cancer. The course ends with "Your Place in the Universe," tracing the history of your own body from the earliest events of the Big Bang.

Module Four: From Atom to Computer
Lecture Nineteen
The Miracle Element

Scope: "Physics in your life" isn't just low-tech physics. Electronic devices, based on semiconductor technology, are ubiquitous in our lives. Increasingly pervasive are digital computers, which include not only our obvious personal computers but, increasingly, any number of "smart" devices from cars to thermostats to phones that all contain tiny computers. Ultimately, the workings of these devices are based in the nature of the atom.

This first lecture in Module Four introduces silicon, the "miracle element" at the heart of modern electronics. Silicon is one of a class of substances called *semiconductors* whose electrical properties lie between those of insulators and normal conductors, such as metals. We have learned to manipulate the electrical properties of semiconductors with exquisite precision, and that manipulation enables the myriad semiconductor devices, from individual transistors to computer chips, that enable our electronic technology.

Outline

I. *Semiconductors* are a class of materials whose ability to conduct electricity lies between that of essentially nonconducting insulators, on the one hand, and the metals and other good conductors, on the other hand.

   A. Although there is a wide range of semiconductor materials, both among the chemical elements and among synthesized compounds, the most important by far is the element silicon. Remarkably, silicon is not some rare, exotic material, but is the second most common element on Earth, after oxygen. Ordinary sand is largely silicon dioxide (SiO_2), as are quartz and most glass.

   B. In the solid state, silicon and other semiconductors form crystal structures in which individual atoms bond to each other by sharing their outermost electrons (covalent bonds; Lecture Thirteen). In this way, semiconductors are like insulators, in which all the outer electrons are locked into the interatomic bonds and are not free to carry electric current. In metals, in contrast, atoms contribute their outermost electrons to a "sea" of free electrons that can carry electric current.

1. Silicon atoms have four outermost electrons that participate in the interatomic bonds. A simplified, two-dimensional model of a silicon crystal shows each atom bonded to its four nearest neighbors. Two electrons join each pair of atoms, so these are double bonds. (The silicon atoms actually bond in three dimensions, with a tetrahedral structure.)

2. At very low temperatures, all the electrons are locked into these bonds, so there are no electrons free to carry electric current. The silicon is an insulator.

3. At normal temperatures, the atoms in the crystal vibrate slightly in the random motion we call, loosely, heat. Occasionally, this thermal energy is enough to free an electron from an interatomic bond. The electron then wanders freely throughout the crystal. Applying a voltage across the silicon crystal produces an electric field that causes the free electrons to move in response, making an electric current. The silicon has become an electrical conductor, albeit a rather poor one.

4. The broken bond, with its electron missing, constitutes a *hole*. Because it's a place where an electron is missing from a structure that was, overall, electrically neutral, the hole acts like a local region of positive electric charge. It's possible for an electron from a nearby bond to leave its bond and take the place of the missing electron, "falling" into the "hole." In this way, the hole can move throughout the crystal. If a voltage is applied to the crystal, electrons will be encouraged to move into nearby holes, with the effect that the holes, acting like positive charges, move through the crystal, making an electric current.

5. There's a big difference, then, between semiconductors and ordinary metallic conductors. In metals, electrons alone carry the electric current. But in a semiconductor, electric current is carried by negative electrons and positive holes. For every electron that's freed from an interatomic bond in pure silicon, a hole is also created. That there are two kinds of current

carriers in semiconductors is what gives these materials their immense usefulness.

C. In pure silicon, there are equal numbers of electrons and holes contributing to the material's electrical conductivity. At room temperature, pure silicon is a modest electrical conductor but not particularly useful or interesting. Pure semiconductors, also called *intrinsic semiconductors*, have little application in mainstream electronic devices, but they do have a few significant uses.

1. Because electrons are freed from interatomic bonds by thermal agitation, the number of electron-hole pairs increases with increasing temperature. Thus, intrinsic semiconductors become better conductors at higher temperatures. Equivalently, the electrical resistance of a piece of intrinsic semiconductor decreases with increasing temperature.

2. A *thermistor* is a piece of intrinsic semiconductor engineered so its temperature-dependent electrical resistance serves to measure temperature. Thermistors are widely used in such everyday devices as fever thermometers, automobile temperature gauges, and household thermostats.

3. Intrinsic semiconductors are also used in some radiation detectors, in which incoming radiation creates electron-hole pairs, resulting in an electrical signal that not only detects the radiation but also provides a measure of its energy.

II. The vast majority of semiconductor applications involve *extrinsic* or *doped* semiconductors—semiconductor materials that have been deliberately "doped" or contaminated with other substances.

A. Atoms of some elements—arsenic, for example—have five outermost electrons that can participate in interatomic bonding, in contrast to silicon's four. If an occasional arsenic atom is introduced into a silicon crystal, the arsenic can "fit" into the crystal structure, bonding like silicon would to its four nearest neighbors. But the fifth electron is extra and can't participate in the bonding. It's free to wander throughout the material and, in response to an applied voltage, to participate in carrying electric current.

1. Even a very low level of such doping can add a great many free electrons, making the doped silicon a much better conductor. Typical doping levels are around 1 arsenic atom for every 10 million silicon atoms. The reason doping is so effective is that there are so few electron-hole pairs formed by thermal agitation in pure silicon. With doping, we get much better conduction and, importantly, precise control over the electrical properties of the doped materials.

2. The vast majority of free charges in the arsenic-doped silicon are electrons. Because electrons carry negative charge, this particular doped silicon is called an *N-type semiconductor*. The *N* doesn't mean the material itself is negatively charged—it's still electrically neutral—but that the charges that are free to move in the material are negative.

B. Other atoms—aluminum, for example—have only three outermost electrons. The occasional aluminum atom can fit into the silicon crystal, but then there aren't enough electrons to complete all the bonds with neighbor atoms. The missing bond is a "hole" into which an electron from a nearby bond can "fall." In this way, the hole can move throughout the crystal. Because it is the absence of an electron, the hole acts like a positive charge.

1. Even a very low level of doping with aluminum or a similar material "floods" the silicon crystal with holes that act as positive charge carriers. Again, the silicon becomes a much better conductor than it would be without doping.

2. The vast majority of free charges in the aluminum-doped silicon are holes. Because they act as positive charges, this doped silicon is called a *P-type semiconductor*. Again, the *P* doesn't mean the material itself is positively charged—it's still electrically neutral—but that the charges in the material that are free to move are positive.

III. Quantum physics provides another way to understand semiconductors.

A. Quantum physics shows that electrons in atoms can have only certain discrete energies. It takes energy to "promote" an electron to a higher level, and energy is released when an electron "falls" from a higher level to a lower.

B. Only two electrons can occupy a given energy level (and the two must have their *spins* in opposite directions). If an energy level is full, then additional electrons can't be promoted or fall into that level.

C. When two atoms combine to make molecules, individual energy levels split to make a more complex structure of molecular energy levels. This is most significant with the outermost levels, whose electrons "feel" and interact with those of the other atom.

D. When multiple atoms combine to make a crystal, each level becomes a nearly continuous band of allowed electron energies, separated by *energy gaps*.

E. The distribution of electrons in the outermost bands determines whether a material is an electrical conductor, an insulator, or a semiconductor.

    1. In a conductor, the highest energy band that electrons normally occupy is only partially full. Electrons can be promoted to adjacent levels in the band with very little energy. That's why the electrons in a conductor respond readily to applied voltages, resulting in the flow of current.

    2. In an insulator, the highest energy occupied band is completely full of electrons. There are no nearby energy levels, and the minimum energy an electron could gain is the large energy of the band gap. Because this is very unlikely, the material has essentially no electrons that are free to move, and it can't conduct electric current.

    3. A semiconductor is similar to an insulator, and at very low temperatures, it becomes an insulator. But the band gap in a semiconductor is very small, so at normal temperatures, some electrons have enough energy to jump the gap, where they then occupy a partially filled band like a conductor's. This is what gives the material its slight electrical conductivity.

    4. Doping a semiconductor—adding the small amounts of impurities that we already saw led to increases in the number of electrons or holes—introduces new energy levels that either donate electrons to the empty band or produce holes by accepting electrons from a nearby filled band. The result is an N-type or a P-type semiconductor, respectively.

IV. It's the existence of two complementary types of semiconductor—P and N—that makes possible nearly all the devices used in semiconductor electronics. Although there are always a few *minority charge carriers* present because of electron-hole pairs created by thermal agitation, the electrical properties of each type of material are dominated by the *majority carriers*—the electrons or holes contributed by the doping material.

Suggested Reading:

Louis A. Bloomfield, *How Things Work: The Physics of Everyday Life*, chapter 21, section 3, pp. 452–453.

Richard Wolfson and Jay M. Pasachoff, *Physics for Scientists and Engineers*, chapter 27, p. 694.

Going Deeper:

Richard Wolfson and Jay M. Pasachoff, *Physics for Scientists and Engineers*, chapter 42, section 3 (a quantum description of semiconductors).

Questions to Consider:

1. Doping silicon with arsenic produces an N-type semiconductor, with each arsenic atom contributing a free electron. How is it that the material remains overall electrically neutral despite all those free electrons?

2. Is a semiconductor more like an insulator or a metallic conductor? Does your answer depend on any other conditions?

Lecture Nineteen—Transcript
The Miracle Element

Lecture Nineteen, "The Miracle Element." This is the first lecture in Module Four, which is entitled, "From Atom to Computer." This is a rather different module than the others. It's very narrowly focused on how, ultimately, computers and other electronic devices work. I'm trying to do a hierarchical progression from individual atoms all the way up to computers, so that you can understand, with no significant gaps, every step in the process that takes us from atoms to building complete computers. Again, you won't be able to build a computer yourself, but you'll have some sense of exactly what's going on inside all of those chips, silicon devices, and keyboards, and whatever else there is that makes up the modern computer.

Now, it's been pretty obvious, I think, so far, but physics in your life is not just the low-tech physics of rolling wheels, bouncing balls, and that sort of thing, but there are a lot of high-tech components involved. The electromagnetic revolution involves high-tech things, and particularly, the electronics revolution. Modern semiconductor electronics is at the heart of so much that goes on in our lives. Increasingly, semiconductor devices, little tiny computers, literally, are built into everything from thermostats to car door opening systems, to car ignition systems, to appliances, and so on. All of these devices are getting smart. They're getting this smarts because they have semiconductor electronics built in to them.

Module Four is about semiconductor electronics—ultimately, how it works, and then, how we build it into practical devices that do other things for us. We begin here with an outline of Module Four, to give us a sense of where we are going. Lecture Nineteen, this lecture, is called "The Miracle Element." It is about silicon, the element behind most of the electronics revolution. Lecture Twenty talks about the 20th century's latest invention— well, with a question mark. I can think of three or four candidates for that award, but certainly, the transistor is one of them, and Lecture Twenty is about transistors.

Lecture Twenty-One, "Building the Electronics Revolution." How do we actually produce semiconductor electronics devices from individual transistors to integrated circuits that today hold one billion or more transistors?

Lecture Twenty-Two, "Circuits—So Logical!" looks at how we actually build up from these transistors to circuits that do very, very simple functions. And it turns out that they really only need to do very, very simple functions that anybody can understand, in order to build up to doing more complicated functions. "Circuits—So Logical!"

Lecture Twenty-Three, "How's Your Memory?" Well, your computer's memory is what I'm asking about there. How is your computer's memory? How big is it, and particularly, how does it work?

Finally, we go back over the whole thing, from atom to computer, and we look at how all the elements of a modern computer fit together, based on what we've seen before.

At the end of each of these lectures, I'm going to have a diagram that starts with a silicon atom, and build us up to where we've gotten to at the end of that lecture. By the very end, it will go from a silicon atom all the way through a series of logical steps to understanding how a computer works.

Let me begin with the "miracle element," silicon. Silicon, for you who would like to know a little chemistry, happens to be atomic number 14. It sits right below carbon, and right above germanium in the Periodic Table. I'm not a chemist, and I'm not going to say much more about that, except to say that silicon is representative of a class of materials called *semiconductors*, which lie somewhere between electrical conductors, and electrical insulators in their electrical conductivity, in their ability to conduct electricity. It isn't just that they are bad conductors or bad insulators, because they conduct a little bit. They work by an entirely different mechanism, and we need to understand that mechanism to understand how semiconductor electronics work.

Now, remarkably—and I think this is almost a miraculous fact—the element that has fueled the electronics revolution is not some rare thing that we have to go to some exotic place on Earth to get, where it happens to be politically unstable, or something, and there's one mine that produces this. Rather, silicon is the second most abundant in the Earth's crust. It is second only to oxygen. It exists largely in the form of silicon dioxide, SiO_2, where it's combined with oxygen and some materials that are basically silicon dioxide, like quartz. Much of the sand on the world's beaches is ground up quartz and glass.

Silicon then is a very, very common material. By the way, you might ask why silicon and oxygen are so common. I will get to that in the very last lecture. It has to do with the nuclear physics going on in the interiors of stars, and what elements are preferentially cooked up. The very final lecture of this whole course, "Your Place in the Universe," will give us a sense of how we, and everything around us on Earth, come from the interiors of stars. That happens to be why silicon is so abundant on Earth. It is the second most abundant material in the crust of the Earth.

Now, semiconductor materials share with other materials a binding together, particularly when they're in the solid state, and that is why devices that used semiconductor electronics are called *solid state* devices, although as it happened, we tended to use that term earlier in the electronics revolution, to contrast them from the earlier vacuum tube devices. Now, almost everything is made up of semiconductors and that is less of an issue.

Silicon bonds itself together into crystalline structure by *covalent bonding*, something I discussed in a previous lecture, one of those methods whereby atoms join together electrically. In this covalent bond, the electrons that join the atoms and that are shared by the individual atoms are basically tied up in this bonding between the atoms; and in principle, and in fact at low temperatures, silicon is exactly like an insulator. It is, in fact, an insulator. In fact, in one sense, semiconductor materials are insulators. They just happen to have some unusual properties. They're qualitatively the same as insulators, but quantitatively different in a way that I'm going to describe.

Now, the key to silicon is its chemistry. Again, I'm not a chemist. Remember that I have mentioned a few times that chemistry is determined by the outermost electrons in atom. Silicon has four outermost electrons that participate in the bonding to other atoms. It's a pretty large number. It's like carbon in that respect. That's because it lies right below carbon in the periodic table of the elements. People have even speculated whether it might be possible that there may be silicon-based life. Carbon's weakness is that it reaches out with these four bonds, and can make very complicated structures.

Silicon bonds itself. Each silicon atom bonds itself through these four shared-electrons to its four nearest neighbors. We are going to be dealing largely with pictures like the one I'm going to be showing here. It's a very simplified two-dimensional model of a silicon crystal. In this model, you see the large red balls are silicon atoms. They're actually positively charged, because the four outer electrons are shown separately. The red are positively charged silicon ions, basically, and you see lines that represent bonds, forces joining each of these to its nearest neighbors.

Because the silicon has four electrons to share, and its neighbor has four electrons to share, each silicon atom, and one of its nearest neighbors, is each sharing two of their electrons. I show those bonds as two lines, with each line having a single electron on it. Each silicon atom is bonded to four nearest neighbors, and there are two electrons participating in each bond. That's the important thing we need to understand.

Now, this is a two-dimensional picture, and I'm going to show it over and over again, to understand how semiconductor's work, but don't think silicon is two-dimensional. Here's a three-dimensional model of a silicon crystal. The structure is actually somewhat complicated. Each atom is joined in a tetrahedral structure, sort of like a pyramid; as if one atom is at the center of a pyramid, and the other four atoms it bonds to are at the vertices of that pyramid. It is therefore quite an interesting, complicated structure.

I borrowed this from a geologist colleague. It is in fact not the structure of silicon, but the structure of diamond; but because silicon and carbon are basically similar, because they occupy the same column in the periodic table, this is also the structure of silicon.

Silicon then is more complex looking than the simplified two-dimensional picture. I'm going to show you. It's a three-dimensional tetrahedral structure, but the basic idea of the physics of what makes semiconductors work will be perfectly clear from the two-dimensional pictures.

Now, at very, very low temperatures, the picture you have on your screen right now is exactly the picture of what a silicon crystal looks like. All of the electrons are participating in these covalent bonds. The outer electrons are all being shared among the atoms, and there are no electrons that are free to move throughout the material and carry an electric current. That is in contrast with a metal. A metal—as opposed to an insulator or semiconductor—when you assemble the atoms in a crystal, the outermost electrons, typically one or two, leave their individual atoms and become completely free to wander throughout the material, forming this sea of free electrons. They can respond to electric fields and carry current. That's why metals are conductors.

This picture I've shown you here is why silicon semiconductors in general, and any insulator, are insulators. The difference between a semiconductor and an insulator in this picture—and I will give you a quantum mechanical explanation at the end of this lecture that will make sense in another way— the energy it takes to dislodge an electron from one of those bonds in a semiconductor is much less than in a normal insulator. That, for example, is the principal difference between the silicon crystal and the diamond crystal. The diamond crystal is a really good insulator. Silicon is not. It's a semiconductor because it takes only a little bit of energy to dislodge one of those electrons.

Where does that energy come from? Well, at normal temperatures—and this will be the subject of the next module, which is on heat-related physics— there is some random jostling of the atoms, electrons, and so on, in other material, and once in awhile, enough energy gets together in one place that it has the possibility to dislodge one of those electrons. At normal room temperature, silicon has an occasional electron discharge from one of these bonds. Not very many, but a few, and there are enough that if I were to apply an electric field to this material, the electron would feel this electric force and it would move. That would constitute an electric current.

Again, thanks to Ben Franklin, because the electric field I've drawn here is pointing to the left, the electron being negative is going to move to the right. There goes. It moves a little bit to the right. It moves a little bit further to the right, and that motion to the right constitutes an electric current.

Here's the picture of a pure piece of silicon semiconductor. At normal room temperatures, it's got an occasional free electron. Not remarkable; and we know that when free electrons are there, they can carry electric currents.

The more interesting thing that's happening is what comes next. Let's focus for a moment on the place where the electron was. There's a bond there. Well, the bond is broken. I have shown it still there, but the bond is partly broken. It's weakened because there is a missing electron. Right next to that electron, though, is another electron. Well, it's not really right next to it, but it's participating in bonding to the same atom, and it feels a slight force from that electric field, also to the right; and although it is in that bond, there's a good probability that it will leap from where it is over into the place where it is not. It won't leap freely through the material, but it may switch which of those two bonds it's on.

In fact, that place where the electron is not—which I have represented here with a little hollow circle—is simply a place where an electron is not. A bond that doesn't have an electron in it, semiconductor people call that a *hole*. It's a place where an electron is not; and the fact is that the electron right next to that hole can move over, and occupy the hole. Then, the bond becomes complete on the right, but not on the left. We would say that the electron "falls into the hole," or would say that the electron and the hole "recombine." What is going to happen now, consequently, is that electron, under the influence of that electric force, is going to jump into that hole, fall into that hole.

There it is. Okay? Well, nothing remarkable there, but here's what has happened. The hole has moved. It's moved to the left, in the direction of the electric field, because absent an electron, it's as if it's a little positive charge. The beautiful thing about semiconductors—and this is the remarkable thing that makes them so useful—is that they have two kinds of things that carry charge. They have electrons and holes. Now, you might think that the hole isn't really real. It's only the absence of an electron. True, but it behaves very differently from the free electrons.

The electrons that are jumping from one bond to another are effectively moving positive charge to the left here, in this picture, and that means that the whole really behaves as a particle. You can calculate its apparent mass.

Now, there is another electron. It's a bond up above, in that horizontal bond, and that's close enough to feel the effect of this hole nearby. Given the electric field that's tending to drive the electrons to the right, it's possible for that electron to jump into the hole, and the hole has moved further to the left. Then, the hole jumps a little further to the left, and the hole jumps a little further to the left, and further to the left; the hole ultimately moves to the left.

The beautiful thing about semiconductors is that they have two kinds of charge carriers, holes and electrons, and it's that fact that gives all the usefulness to semiconductors. It's the manipulation of that phenomenon— holes and electrons carrying electric current—that makes the semiconductor revolution possible.

In fact, in the so-called *intrinsic* silicon—this is pure silicon—what happens when an electron is free is that the hole is created, and in fact, the electron-hole pair is created. I will say a lot more as I go on about electrons and holes, and electron-hole pairs. The point is though, there's a big difference between semiconductors and ordinary metallic conductors. In metals, there are only electrons carrying the current. In semiconductors, holes also carry the current.

Now, in pure silicon—as you can see the picture, where I show an electron-hole pair being created—for every electron that's free, there's also a hole, so electrons and holes occur in pure silicon in exactly the same numbers. The conduction is equivalent by each of them.

Pure semiconductors aren't terribly useful. I will show you a couple of examples of where they are used, but they themselves are not that useful in the electronics revolution. Let me just give you one example of what we do use a pure semiconductor for, however. Then, I will move on to tell you what we really do.

Let me see if I'm sick. No, I'm okay. How do I know? Because your modern thermometer doesn't contain mercury, a horrible, toxic substance that expands, although its battery may have some mercury. It has a tiny little place in its tip where there is a tiny little chunk of an intrinsic semiconductor, probably silicon. An "intrinsic" semiconductor means that the semiconductor is really pure. That's the only thing in it.

A piece of pure semiconductor is called a *thermistor*. I have over here a thermistor. You can just barely see it. It's this little tiny device in a little glass bead, and there are two wires coming out of it. The beautiful thing about a thermistor is that these electron-hole pairs, remember, are produced by agitation due to thermal energy. The hotter it gets, the more electron-hole pairs ought to be created. Therefore, a piece of pure intrinsic semiconductor becomes a very nice temperature sensor. Right now, this thing is sitting at room temperature, and the meter is measuring its electrical resistance. It is 9.26 thousand ohms. An ohm as a unit of electrical resistance, it's the amount of resistance to cause one volt to force a current of one amp to flow. One amp would cause a current of much less than one amp to flow, in this case, about one 10,000ths.

Now, I'm going to put my fingers on this thermistor, this piece of intrinsic semiconductor. That's going to make more electron-hole pairs, and you can see that driving the resistance down. I'm going to take that thermistor, and I'm going to dip it in ice water. You can see the resistance going up. It's now off scale. I have to press a different button. It's now 22,000 ohms. A piece of intrinsic semiconductor, although it's not what we use making most of the electronics revolution, becomes a very useful temperature sensor.

Another thing intrinsic semiconductors are used for is radiation detectors, detecting nuclear radiation. That works because when high-energy radiation comes in, it produces electron-hole pairs, and we can actually measure not only the fact that radiation is there; but by determining the energy given to these electrons and holes, we can determine the energy of the radiation, so some sophisticated radiation detectors used intrinsic semiconductors as well.

That's not the big deal, though. The big deal is that we can modify the properties of semiconductors. The whole electronics revolution is based on our ability to do that. What we do is intentionally contaminate pure, pure silicon. We have to start with really pure silicon. We contaminate it with low levels of contaminants. The levels of contaminant we are talking about typically vary, depending on what the application is. They are typically on the order of something like one impure atom in about 10 million silicon atoms. That has to be controlled very carefully, but that control gives us immense power over the electrical properties of the material.

Here's a picture of silicon that's been doped by arsenic. Arsenic is not such a nice material, and that's why some of the products used in making semiconductors are terribly toxic. Arsine gas is something you don't want to be near, but it's available in semiconductor manufacturing facilities. It's pretty bad stuff.

I've shown you the arsenic atom. It's bigger than the silicon atoms, and the thing about arsenic is that it has five electrons in its outermost layer. It is pentavalent. It has five electrons that it would like to contribute to bonding. It doesn't really bond, the family silicon does, but with the arsenic at the level of one in every 10 million for silicon, it will do the best it can to fit into the crystal silicon structure. The problem is that it will have one extra electron, and that won't fit into that bonding structure, which requires four bonds to your nearest neighbors. It's not what arsenic wants to do, but it will nestle into that crystal structure, and try to fit in there as an impurity.

The problem is, it gives up this free electron, and it can't participate in the bonding. That electron becomes free, like the electrons in metal, to conduct electric current. That electron, even at the level of one arsenic atom in every 10 million silicon atoms—those electrons contributed by that small amount of arsenic are enough to greatly swamp the electron-hole pairs that are created intrinsically, and they become the dominant charge carriers in this material. There's a free electron in there, and that makes the material a much, much, much better conductor that would be otherwise.

We call those electrons in this material the *majority charge carriers*. There are still some electrons that have been released by the thermal process. They're also majority carriers, but because they mix in with these, they aren't distinguishable, and there are a few holes. They're called *minority carriers*, but the vast number of charge carriers in this material, this arsenic-doped silicon, are electrons, and for that reason, the arsenic-doped silicon is called an *N-type semiconductor*, "N" for negative.

Now, be very clear on this. That doesn't mean that the material is negatively charged. The arsenic gave up this extra free electron, but it also had an extra proton in its nucleus, so that the material is still electrically neutral. "N-type" means not that the material is negative, but that the things that are in the material that are free to carry electric current are negative, and they are electrons.

Now, those are not the only types of atoms we could dope semiconductor with. We can also dope a semiconductor with atoms that have fewer electrons. An example of that is aluminum. Here's aluminum-doped silicon. I have shown the aluminum atom to be slightly smaller. It is still a positive ion in there. It has three outermost electrons to contribute to this bonding, so it nestles into this crystal structure, which again it can do when we're doping at the level of one part in 10 million, or something.

It nestles into that crystal structure, but there's a bond that doesn't have an electron in it. Well, we know about the absence of an electron in semiconductors. That's a hole. This aluminum-doped silicon produces holes that are not matched by electrons. It does not make electron-hole pairs. It simply put a few places in the structure where there are not electrons, and they behave like holes do. I showed you that before. The hole can move through the material in response to an electric field. Even at very low levels of doping—again, one part in 10 million, or something like that—these holes dominate the electrical conduction in this material, and therefore, we call that a *P-type semiconductor*. Again, it's not positively charged. It is still electrically neutral, but the things that are free to move in it are holes. They're positively charged.

Now, you might say, "Yeah, but are they really electrons that move?" Yes, but they aren't moving freely. They're jumping from one bond to another. It's basically the hole—positive thing—which is free in this place to move throughout the material.

The key here is that we have two different types of semiconductors. We have P-type, and N-type. Putting those two together gives us a host of wonderful responses that determine how we build our electronic devices.

Now, before we go on and start building these devices, I want to give you just a little more insight into these P- and N-type materials, as well as semiconductors versus insulators and conductors, and a quantum mechanical sense. I haven't said much about quantum mechanics in this course, or quantum physics. You may know enough to know, though, that quantum means "discrete," little discrete hunks, and one of the implications of quantum physics is that the electrons in atoms cannot just have any energy, or being just any orbit around the nucleus. Only certain orbits are allowed.

That fact, which you probably learned somewhere in high school science, maybe, is the key to understanding semiconductors from a quantum physics point of view. Here's a picture of a quantized atom. Again, this is this crude solar system model and it really doesn't correctly represent what an atom looks like in quantum physics. The electrons are, again, in a sort of statistical cloud, but this picture will help us understand what's going on. I'm showing two levels of orbits; and in fact, typically, two electrons can occupy exactly the same state. Well, not quite exactly, because one of them has its spin pointing down, and one pointing up. I'm not going to go into that, but the fact is that at most, two electrons can occupy what you would think of as an orbit.

I have shown three orbits are, two of which occupied. They're full, in fact, with two electrons each, and one is unoccupied. If I drew a diagram of the energy associated with those orbits, it might look something like this. On the vertical direction is energy, and the horizontal direction in this picture is meaningless. Think of those as little shelves, different levels where electrons can live. This particular atom now has two electrons in the lowest level, and two electrons in the next level, and no electrons in the third level.

It's possible to promote an electron upward, and I've made it happen here. Let me do that again. I made it happen in both pictures in the actual physical picture of the atom, and in the energy diagram, one of the electrons in that outermost, occupied ring is going to jump upward. It takes some energy to do that, that discrete amount of energy. That's what causes spectra, and all kinds of other things, from atoms.

Well, that's all well and good, but the semiconductor electronics is not about individual atoms. It's about atoms coming together in the solid state to make dense crystal structures. We want to know what happens to this quantization of electron energy levels, to the fact that there are only discrete levels when you begin to bring atoms together. Let me show you what happens.

Here are a couple of atoms, and if they are far apart, each of those atoms is exactly the same energy level structure that I showed before. Electrons occupying the upper atom, and electrons occupying the lower atom don't know anything about what the other atom is doing, so they are happily occupying their levels as if they were the only atoms of that sort in the Universe. However, bring the atoms together, and in this energy level diagram, as we saw before; each atom does. Bring the atoms closed together, and now they're close enough together that each atom "knows," through the medium of their interacting shared electrons, what's going on in the other atom.

An electron that was happily in its level in one atom, and an electron that was happily in the same level in the other end, when they were far part—now, they can't be in the same level. Consequently, those levels split slightly, into two levels separated slightly in energy. This is exactly the kind of thing chemists who are trying to understand chemistry from first principles—like the model—they want to understand what happens if you put the two atoms together, and what becomes the new, more complicated energy levels in the molecules, because what I have here is a diatomic molecule—two atoms joined together to make a molecule.

The electrons deep down inside—the so-called lowest energy electrons that are tightly bound to the atomic nucleus—their levels are not affected much by this. There, again, they're still largely isolated by the outermost electrons from this process of joining. The outermost electrons though, the ones that participate in the chemical bonding, really understand now that they're part of a two-atom system, and they really feel these different levels. I showed the upper levels splitting more dramatically, therefore, than the lower levels.

This is molecular physics, or the beginnings of chemistry, but it certainly isn't solid state, or what we now call condensed matter physics. It isn't a structure as complicated as this is. However, what if we join lots of atoms? Well, if we join lots of atoms to make a solid structure—and here's a representation of the crystal. This one only has 16 atoms in it, and a real one has a 10 with 20-something zeros after it, in typical microscopic sizes.

Those levels become so close together, and there are so many atoms, that we get, effectively, where each level is a whole band of allowed energies—even though in principle, if the crystal is finite in size, which every real crystal is, they're actually discrete individual levels they're so close together that they become completely indistinguishable; and again, the further you are up in energy, the wider those bands are.

Those bands of allowed energies are separated by gas, and it is the band theory of semiconductors that is ultimately the way of understanding them on a mechanical level. Let me emphasize again that this is beginning to look sophisticated and complicated, but it is nothing but a manifestation of the fact that individual electrons and atoms can only have certain discrete energy levels, and its complication of that fact that results as we bring more and more atoms together. Bring two atoms together, and those levels split in two. Bring zillions of atoms together, and those atoms split into zillions of closely spaced levels that form these bands. There are still gaps though, and they correspond to the gaps between individual levels in the original atoms.

What happens here? Let's talk about conductors, insulators, and semiconductors, in the context of this theory. Here's what makes a conductor. A conductor is a material—and I'm showing you here only the outermost two energy bands that would have electrons in them. There are a whole host of higher bands, but they typically don't have any electrons in them. In a conductor, the next highest band is completely full, but if the outermost band contains any electrons, it only has some of its levels occupied. In this picture, I'm showing the band in gray, as I did in the previous picture, but I'm showing the occupied state, the occupied levels, in blue. That means that there are electrons in those energies.

Because there are only some electrons in the upper band, the energy difference between the highest energy electrons at the top of that band, and the next level—remember that these levels are so close that they form, essentially, continuous band—there's almost nothing. It therefore takes almost no energy to promote an electron from that highest occupied level into the unoccupied levels. Once the electron is in an unoccupied level, it's free to move throughout the material, because there are additional unoccupied levels right next to it. You can apply an electric field, and you can get more energy.

That's the point. If an electron can respond to a tiny bit of energy, then it is free to move throughout the material. That's how a conductor works in band theory. It has a filled band, and a so-called *conduction band*, which is partially filled.

How does an insulator work? Well, an insulator has a different structure. It has a filled band, and then, there's a big gap. There it is. There's a large gap before there is an unfilled band. It's the empty band. Sometimes, it is called the *valence band*. It is unfilled. That band has no electrons in it. In order to promote one of the electrons from the filled band to the empty band, you have to apply a very large amount of energy. You could, in principle, apply enough electric field, pull those electrons out of the bonds and into the insulator, but it would take a huge electric field. That's why insulators are insulators.

Finally, what about semiconductors? Well, they are, as I said, essentially insulators. The difference is that the band gap is much smaller. They have a very small gap; and at absolutes of temperature, they are insulators.

There's no way to get across that gap. But at even any reasonable temperature, thermal energy is enough to promote a few electrons into the empty band. Those are the ones that I showed before as being free, and those of the ones that become free to move throughout the material. They are the conduction electrons, and now, the material has a structure like the structure of a conductor.

In an N-type semiconductor, what we do is introduce a new energy level associated with the impurities, called the *donor level*, and that can easily give lots of electrons to this unfilled band to become a conductor.

In the P-type materials, it's the same thing. We put an empty band down below. We can promote some electrons into that band, and we again get conduction, now by holes, because of that empty unfilled space down below.

That's an explanation of semiconductors on the basis of these bands. Let me wrap up with the picture I'm going to show you at the end of every lecture in this model. We started out with silicon atoms, and built silicon atoms into crystals. The crystals had electrons and holes as possible current carriers, and so, we developed out of that N- and P-type conductors. From that, we're going to build the whole semiconductor revolution.

Lecture Twenty
The Twentieth Century's Greatest Invention?

Scope: The 20th century saw unprecedented technological advances, with major inventions in areas ranging from medicine to agriculture, to space flight, to optics, to mechanics, to electronics. Near the top of virtually every list of the most important inventions of the 20th century is the transistor, a tiny semiconductor device at the heart of every electronic gadget from the simplest radio to the most complex supercomputer. The transistor's secret is that it lets one electrical circuit control another. The transistor, and its simpler cousin, the diode, are created in the conjunction of the two types of semiconductor, N and P, and depend for their operation on the remarkable properties of the PN junction.

Outline

I. A *PN junction* forms at the interface where a piece of P-type semiconductor is joined to a piece of N-type semiconductor.

A. In the common process of *diffusion*, a material tends to move from a region of high concentration to a region of lower concentration. Open a bottle of perfume, for example, and perfume molecules diffuse from the bottle and eventually reach the far corners of the room. There's no magic here; it's just that there are more molecules in the region of high concentration, and they're moving in random directions. Thus, there are more molecules moving away from the region of high concentration than there are moving toward that region.

B. When P and N semiconductors are first brought together, there is a high concentration of holes in the P-type material and a high concentration of electrons in the N-type. Thus, each type of charge carrier diffuses across the junction. Electrons diffusing into the P-type material "fall" into the holes they find there, reducing the number of free charges just on the P side of the junction. Holes diffusing into the N-type material encounter electrons; again, electrons and holes recombine. The result is a dearth of free charges in the region near the junction, and as a result, the junction region becomes a very poor conductor and a barrier to the flow of electric current. Because it's been depleted of free charges, the region around the junction is called the *depletion region.*

C. Because the P- and N-type semiconductors were initially electrically neutral, diffusion of positive holes into the N-type material gives that material a net positive charge in the region of the junction. Similarly, electrons diffusing into the P-type material give it a slight negative charge. Eventually, enough charge builds up to repel additional electrons trying to diffuse into the P-type material and holes into the N-type material. As a result, a balance is established between diffusion and electric repulsion. This balance develops almost immediately after the PN junction forms.

D. If a battery is connected to the P and N materials, with the battery's positive terminal to the N-type material, the effect is to widen the depletion region and strengthen the electric field, making the junction an even poorer conductor. This condition is called *reverse bias.*

E. But connect the battery with its positive terminal to the P-type material, and the depletion region shrinks. The junction becomes a decent conductor, and current can flow through the device from P to N.

F. All this sounds complicated, but it boils down to one simple fact: A PN junction conducts electricity in one direction but not the other. It conducts when its P-type side is made more positive and the N-type side is made negative. It doesn't conduct in the opposite case. That's all we need to know from now on.

II. Joining P- and N-type materials makes a *diode*, a useful device in its own right.

A. Diodes act as one-way valves for electric current and are, therefore, used to convert alternating current (AC) to direct current (DC). Because virtually all electronic equipment uses DC, diodes are ubiquitous in electronic devices powered by regular AC household wiring.

B. In a diode that's conducting current, electrons and holes are continually recombining at the junction. Energy is released when an electron "falls" into a hole. In *light-emitting diodes* (*LEDs*), the semiconductor materials are engineered to have an energy band gap such that the energy is released as visible (or sometimes infrared) light. Most of the small red and green lights used as indicators in electrical equipment are LEDs, because they're more energy efficient and far longer lasting than traditional lightbulbs. High-intensity LEDs are now available in flashlights and other applications, and it is only a matter of time before they replace lightbulbs in most everyday lighting applications.

C. The reverse process can occur, too. In a *photovoltaic (PV) cell*, incoming light energy dislodges electrons, creating electron-hole pairs (recall Lecture Sixteen). When these pairs form near a PN junction, electrons move into the N-type material and holes into the P-type material under the influence of the electric field established at the junction. As a result, the N-type material becomes negatively charged and the P-type material becomes positively charged. Connecting an external circuit allows current to flow, delivering electrical energy. Thus, the PV cell converts solar energy into electrical energy. One trick in making efficient PV cells is to utilize different materials whose band gaps correspond to different energies in the solar spectrum.

III. A *transistor* is a device containing two PN junctions, which allows one electric circuit to control another. There are many types of transistors, but the easiest to understand is the *field-effect transistor (FET)*, widely used in computers and other electronic devices. The particular transistor I'll describe is a *metal-oxide-semiconductor field-effect transistor (MOSFET)*.

A. In a common type of FET, two pieces of N-type semiconductor are embedded in a block of P-type material.

B. If a battery is connected via the wires attached to the N-type pieces, no matter which to the positive battery terminal and which to the negative, no current flows. That's because there are two PN junctions where the N-type pieces join the P-type, and no matter how the battery is connected, one of these will always be connected in the way that blocks current.

C. There's more to the transistor. A thin layer of insulating material (usually silicon dioxide, made by exposing the silicon block to oxygen) is formed over the P-type "channel" between the N-type pieces, and a layer of metal is coated on top of the insulator. If the metal (called the *gate*) is connected to the positive terminal of a battery, it becomes positively charged, repelling holes and attracting electrons into the material below it. Even though the material has been doped to be P-type, the influx of electrons makes it temporarily N-type. Now, the PN junctions are temporarily gone, and there's no barrier to current flow between the two N-type pieces. The transistor has become a conductor.

1. How good a conductor it is can be controlled by the amount of charge on the gate, which is determined by the voltage applied to the gate. Thus, the gate voltage controls the current through the transistor. A very weak electrical signal at the gate can control a much larger one, making the transistor an *amplifier*. Amplifiers are used throughout audio and video electronics to boost the level of electrical signals from microphones, CD pickups, tape heads, and so forth.

2. The transistor can also be operated in a mode where the gate voltage swings between two extremes, making the device behave like a switch that is either on or off. In this mode, the transistor forms the heart of digital electronic devices.

3. It's possible to build a complementary transistor, with P-type material embedded in an N-type block. The use of complementary transistors enables energy-efficient circuitry ranging from laptop computers to high-power audio amplifiers.

Suggested Reading:

W. Thomas Griffith, *The Physics of Everyday Phenomena: A Conceptual Introduction to Physics*, chapter 21, section 3, pp. 453–455 (Griffith describes a different type of transistor).

Richard Wolfson and Jay M. Pasachoff, *Physics for Scientists and Engineers*, chapter 27, pp. 695–696.

Going Deeper:

Paul Horowitz and Winfield Hill, *The Art of Electronics*, chapter 2.

Neil Storey, *Electronics: A Systems Approach*, chapter 5.

Questions to Consider:

1. How is a diode like a one-way street?

2. In a lightbulb, current flowing through the bulb's filament heats the filament until it glows. In a light-emitting diode (LED), current flowing through the device results in electrons and holes recombining, producing visible light in the process. Why might an LED be more efficient than a lightbulb in converting electrical energy to light?

Lecture Twenty—Transcript
The Twentieth Century's Greatest Invention?

Welcome to Lecture Twenty, "The Twentieth Century's Greatest Invention?" with a question mark after it, because there are many candidates for the 20th century's greatest invention. As a physicist, I tend to think of things like the one in this lecture, the transistor, but I also think of things like lasers. I think of things like things like the airplane. I think of the computer, although the computer depends largely, but not entirely, on the transistor. Others might argue that it's the "green revolution," which has enabled us to feed a population of a world with six billion or more people.

Others might argue that it would be genetically modified organisms, cures for certain diseases, antibiotics, and so on. There are many candidates for the 20th century's greatest invention, but surely the transistor ranks up there among them. The transistor is that the heart of all modern electronics. It's the single element that goes the real work of modern semiconductor based electronics. The transistor has some antecedents in the vacuum tube, electromechanical relays, and other things—more on that in the next lecture, when we actually talk about how we build this thing. For now, I want to show you how at least one simple kind of transistor works. It's a kind of transistor that's widely used in electronic circuits.

This is going to get a little bit technical. This is probably the most detailed physics you're going to see of my lectures, in some sense. It's not mathematical, but some pretty subtle things are going on, so try to follow, and if you have to back up and play it over again, by all means do that. You will come out at the end of this lecture understanding exactly how a transistor works.

Before I talk about transistors, I have to talk about a general physical process that occurs that sometimes people think of as rather mysterious, or get confused about. It's really very simple. It's a process called *diffusion*. Sometimes we say, "Nature abhors a vacuum." Materials expand to fill a vacuum.

I'll give you another example. What if I get a little bottle of perfume here and I unscrewed it? Unfortunately, scent doesn't go through onto the DVD that you are watching, so this won't work for you, but other people in this room would eventually smell that perfume. They would even smell it if the wind were blowing the other way, because diffusion is a process that simply does this. If I have a region where there is a high concentration of something—air molecules, perfume molecules, electrons, whatever—that high concentration tends, over time, to spread out and even out the concentration.

In particular, the things of which there are a high concentration—air molecules, electrons, whatever they are—tend to move into regions of lower concentration. You can even see this if you're driving on the highway and it's very crowded in one lane, and another lane is not so crowded. There is a kind of instinctive movement into that other lane. That is kind of a process of diffusion.

How does diffusion work? I have a simple picture that suggests what's going on with diffusion. Maybe these are air molecules, maybe they're electrons, or maybe they are cars. I don't care what they are, but on the left, there's a high concentration of them. They're moving in random directions. I showed them all the same speed, but they could be random speeds, too.

On the right, I have a region where the concentration is lower. If those molecules, whatever, are moving, again, in random directions—of course, there would be zillions of electrons or molecules, not just a few, but here's the point—if I imagine an imaginary dividing line between these two regions of high and low concentration, and ask the question, "What's the rate at which particles are moving from left to right across that barrier—or not barrier, but imaginary dividing line—versus the rate at which they're moving from right to left?"

Now, the only thing that depends on is how many particles there are to the left that happen to be moving toward the right, and how many particles are to right, that happen to be moving toward the left. Well, there are more particles at the left, moving in random directions, so there are more particles of the left that are moving toward the right, then there are particles on the right that are moving toward the left.

As a result, there is a net flow of particles across that imaginary line, driven by the fact that there are simply more particles on the left. That's the process of diffusion. Diffusion is a process where a high concentration of particles in high concentration tend, really, to spread out into regions of low concentration, simply because a few draw any line between a region of high and low concentration, there are more particles moving one way across that line, then there are the other way, simply because there are more particles on one side of the line than there are on the other. Eventually, if things even out, there still particles moving randomly back and forth, but when the concentrations become equal, then there are equal numbers moving either way, and there's no net flow of particles.

If I come back to this situation after awhile then, diffusion has occurred, and particle concentration is more or less even now. Nature abhors a vacuum because the random motions of particles take them from regions of higher concentration to lower concentration. That's a general fact of physics and chemistry. It has something to do with semiconductors, but it also has to do with the spread of pollutants, with the spread of voters, with the spread of radioactive materials, and with all kinds of other things besides semiconductors.

We're going to deal with it in the case of semiconductors. Here's the situation. The key idea to understanding semiconductor physics is to understand, first of all, that we have two different types of semiconductors: P-type—in which the charge carriers are positive—they are holes. N-type— in which the charge carrier is negative—are electrons. I want to emphasize one other thing I told you when I first introduced those ideas. "N" and "P" stand for negative charge carriers, but they don't mean that the material in bulk is negatively charged.

Give me a random piece of N-type material, isolated by itself, and it has no net electric charge. It has as much positive charge in it as it does negative, but it's just that the negative charges are free to move. Give me a random piece of P-type material, all by itself, and it's electrically neutral. It has as much positive charge as it does negative. It's just that the positive charge— the holes—are the things that are free to move.

Keep that in mind, because in a little bit, we are going to start making these things be charged in ways that may seem a little bit counterintuitive. If you want to think that "P" means that it is positively charged, it doesn't. P means that for the dominant charge carriers, the majority carriers, the holes, are positive. N means the majority carriers are negative electrons. The key to semiconductor physics is to put P- and N-type materials together, to join them at a junction between these two very different kinds of materials, which were made different by that process of doping with pentavalent materials like arsenic, or trivalent materials, like aluminum, that didn't quite fit into structure and therefore gave us additional free electrons in the case of the arsenic, or additional holes in the case of the aluminum, therefore making us N- or P-type materials.

What we want to do is examining detail what happens when we join P- and N-type materials. That is called a *PN junction*. Here I show a hunk of P-type material on the right, a hunk of N-type material on the left. They are, as I just said, electrically neutral, and I'm going to bring them together to form a junction.

Now, I have an imaginary line. Well, it's actually not so imaginary. It's a real line. This is a junction between those materials, and of course this is a two-dimensional picture of something that's actually happening in three dimensions. However, here's the issue: On the left, there's a high concentration of holes, and holes are mobile in the P-type materials. They are free to move. On the right, there's a high concentration of electrons, so that when I put this structure together, when I first assemble a PN junction, holes are going to diffuse from the left side of the junction, from the P side of the junction, to the right side of the junction, the N side of the junction. Electrons are going to do the opposite. Electrons will diffuse from the N side of the junction, where they are in high concentration, to the P side of the junction.

Now, I have electrons in the P-type material. This is why I didn't want you to think that "P" meant, "positively charged." It means that the majority carriers are holes; because there was a high concentration of holes, they have diffused to the right, and entered the N-type material.

Now, the N-type material is full of free electrons. Well, all of a sudden, there are a lot of holes in there. There are a lot of places where there aren't electrons. If one of those free electrons stumbles upon a hole, it's actually energetically favorable. It's like something falling down. It can fall into the hole, and the two of them will recombine.

Here's the process of recombination. If I have a hole, and I have an electron, and they get together, they join, and there's nothing left. I mean, yes, the electron is still there. It's now part of the bonding structure, and isn't participating in this electrical conduction. As far as electrical conduction is concerned, this process of recombination—an electron falling into a hole; a hole and an electron coming together—removes two charge carriers from the system. The electron is still there, but the hole isn't there, because the electron is now occupying the bond. The electron hasn't disappeared from the Universe, but it's no longer playing any role in electrical conduction.

What happens when we have just diffusion process at the PN junction is that holes are diffused into the N-type material, and electrons have diffused into the P-type material. Then, recombination occurs, and consequently, right around the junction, there's a region that is depleted of charges. There are very few free charges, because holes have migrated to the right, and combined with electrons, and electrons have migrated to the left, and recombined with holes. The junction region becomes depleted of charge areas, free charges that are free to move.

Free charges that are free to move are what make something a conductor. Therefore, the junction region, if I just form a PN junction, becomes a poor conductor. It makes this block of material that with each individual part, had I tried to determine its electrical conductivity, tried to pass current through it—either one of them would have done a fairly decent job of passing current. Not a great job, because it is a semiconductor, but a better job than a pure semiconductor, because these are both doped to have high concentrations of either electrons on the right—in the N material—or holes on the left—in the P material. Now we've got a region in this block of material that is depleted of charge carriers, so it's very difficult for current to flow through this device.

Great. I've taken all of this complicated technology and have made something that doesn't let electric current flow. There's more to it than that, though.

Even though the electrons that diffused from the right hand side, from the N-type region into the positive region are no longer free to conduct, they're still there. There is still an excess of electrons, and so the right hand edge of the P-type material near the edge of the junction in fact has a net negative charge. Again, that's why I told you not think of "P" as meaning that it is positively charged. It means that normally, at least, the majority charge carriers are holes. In this case though, the right hand edge of the P-type material is carrying a negative charge.

Similarly, holes that are fused into the N-type material, bringing with them positive charge, we have combined with electrons, so that the holes are not there any more. However, the dearth of electrons that they represent is still there as an imbalance, and there is a net positive charge on the right hand side of the junction, even though those charges are not free charges that are free to move and carry current. That kind of distribution sets up something we know about from earlier lectures. It sets up an electric field that points in that direction from the positive toward the negative, and if a positive charge got into that region, it would want to be accelerated that way.

Now, you might ask, "Why doesn't this diffusion process keep happening until the N- and P-type materials are completely gone?" Well, the electric field exerts forces on positive charges that are in the direction of the field— that is, to the left—and that means that as this electric field builds up, as more and more holes migrate to the right, and more and more electrons migrate to the left, this electric field gets stronger and stronger, and it makes it harder and harder for those holes and electrons to continue diffusing. The holes are having to buck the electric field, going against it, and they feel a force in the direction of it, so it gets harder for them.

The electrons would have to move in the direction of the field, but electrons want to move in the direction opposite the field, because they are negative. Consequently, both kinds of charge carriers now have trouble diffusing, and eventually what happens—and it happens very, very quickly, almost instantaneously, when I first build a PN junction—the diffusion happens, and the diffusion builds up this field, and electric field stops the diffusion.

Now, at a microscopic level, the particles are still moving back and forth, but we soon, in a tiny fraction of a second, reach an equilibrium in which there are as many carriers moving across the junction one way as the other, and the electric field is basically halted, and balks the diffusion process. We end up sitting here with—although it might be microscopic, motions are still occurring—no bulk flow of particles. There's no more diffusion. There's an electric field established, and we are all well and good.

That's a PN junction. Now, here comes the fun part. We want to do something with this device. We want to connect it to external circuits. We want, for example, to connect a battery to it. Suppose I connect a battery to it like this picture shows, with the positive end of the battery, the positive terminal of the battery, the terminal of the battery that wants to drive positive charge to it. I've connected that to the N-type material, the N-type side.

I've connected to negative terminal of the battery to the P-type material. Remember what the N-type material is full of. "N" means "negative." It's full of negative charges. It's not negatively charged. In fact, this particular N-type has a net positive charge near the junction, but it still has mostly electrons, N-type charge carriers. They are therefore attracted by this terminal of the battery, and they move away, out of the N-type material, and the effect of that is to make the depletion region even bigger.

Similarly, the holes in the P-type material are attracted toward the negative terminal of the battery. They can really get there, but they get toward the negative end of the P-type material. That causes the depletion region to be even broader, because you have sucked holes away from the junction on the left. You have sucked electrons on the right, and the depletion region is even bigger.

If you connect a battery this way then to the PN junction, almost no current flows. It's a very, very poor conductor of electricity. We've made things even worse. We've put a lot of sophisticated technology into this and nothing happens.

However, if I connect the battery the other way, connect the positive to the P, and the negative to the N, exactly the opposite happens. Holes are driven into that region, and electrons are driven into that region, because they are repelled from the negative terminal and are attracted—and holes are repelled from the positive terminal. The depletion region once again—it was a depletion region—the region around the junction fills up with free carriers that are free to carry current, and the whole block of the PN-type material becomes an electrical conductor. Electrical current flows happily around the circuit.

That's what a PN junction fundamentally does. It blocks the flow of electric current in one direction, and allows the flow of electric current in the other direction. The detailed physics of this is in what's happening at the junction, with the balance of diffusion, then this electric field, and then additional electrical effects supplied by this battery. This is called *forward bias*. We have the battery's positive connected to the P-type material, and current can flow. The other condition is called *reverse bias*, where the battery is connected the wrong way, and current cannot flow.

From this point on, you can forget that fundamental physics, but you know it, and you know how we built P and N from silicon and doping, and you know all the way back to the beginning, the atom, but from now on, we're going to build on the fact that we have PN junction, and this is what they do.

They are essentially one-way valves of electricity. You can have a current flowing from P to N where the device is perfectly happy to do that. It's a one-way valve, and you cannot have current flowing in the other direction. That's all you need to know about a PN junction.

Now, there are some subtleties here. There's a tiny bit of current flowing in the forward direction. The current flowing in the reverse direction is not a perfect conductor, but it's a pretty darn good conductor. The forward direction is a very poor conductor in the negative direction. If you connect one of these things to an ohmmeter, like I had in the last lecture, you would find that it had a very high resistance in one direction, and a very low resistance in the other.

Now, this device, a PN junction, is not quite a transistor yet. However, a PN junction is useful in its own right. For example, almost all of our electronic devices require direct current, steady current that flows in one direction for their power sources. Whenever that failed, and you had variations at the 60-cycle AC power line frequency, all of your stereo would go, "Buh!" It would make a horrible 60-cycle noise, and I've seen that happen when certain devices failed.

The fundamental device that turns alternating current to direct current is the PN junction diode. It's called a *diode*, because it has two elements to it, just two places where you connect wires to it, the ends of the P- and N-type materials. That's a diode, and diodes are used for so-called *rectification* purposes, turning alternating current into direct current. Every electronic device you have has lots of diodes in it; well, typically four. Sometimes only one will do it for that purpose.

That's one use for these simple PN junctions. Another use, which is increasingly important, is that in some semiconductors—not silicon itself, but I other semiconductors—by modifying the different combinations of elements according to make a semiconductor, you can actually vary that band gap that I talked about last time. When you vary that band gap, the not so large gap between the outermost band and the filled band in the semiconductor—the gap across which some electrons can jump—you can vary the width of that gap, and when electrons and holes recombine, they give off an amount of energy equal to that band gap, and that energy comes out as a little burst of electromagnetic radiation called a *photon*. If you engineer that band gap just right, that photon will be in the visible region of the spectrum, and it will make visible light.

One of the most important devices we have today then is the light-emitting diode, and the light-emitting diode is simply a PN junction diode. It's not silicon, but it is engineered with other materials to give off light of particular colors. You can engineer it for different colors, and I have here—we will take a close-up shot at it—a little circuit board that you can build simple electronic circuits on. I have on this end of it, on the right, a couple of diodes, a red LED and a green LED. I also have a little resistor, a little tiny device that has significant electrical resistance. That's because if I connect the battery right across the diode, in the reverse direction, nothing would happen. In the forward direction, large amount of current would flow, and the diode would be destroyed. You always, therefore, have to put a resistor in series with the diode.

I'm going to connect the negative terminal to that end of the LED. The LEDs are connected together, and the other ends are connected together through the resistor. Both of these LEDs are therefore in the circuit. Right now, the green LED lights, very happily. The red LED is off. The reason is that I connected these two light-emitting diodes back to back, reverse direction, so that one of them has its P connected to the other one's N, so that when I try to make current flow through them, it flows through one, but not the other.

Now, if I reverse the polarity of the battery—here's the negative terminal of the battery, and here's the positive terminal—if I reverse that, the other diode becomes forward biased, and lights. Back and forth. The green one is lighting, the red one is lighting. The green one is lighting, the red one is lighting. Right now, the red one is forward-biased, and the green one is reverse-biased. Now, the green one is forward-biased, and red one is reverse-biased. Those are light-emitting diodes.

Light-emitting diodes also see applications in devices that you find all over the place these days. That is, in these simple seven-segment displays that I have at the left-hand end of this board. There's a little display. It has seven little segments that together can make any of the numbers, and those seven segments are made up of light-emitting diodes. Here, I will light one of those segments. I'm not going to light them all at once. In a real computer circuit, or digital readout of any kind, electronic logic circuits such as we will be discussing in subsequent lectures will determine which combinations of these have to light, so that if it were a "three," it would be the top one, one of the side ones, and side ones on the other side, and the bottom and middle one would all light in a "3," for example.

If I move around which one I'm connecting to, for example, I can light different segments. There's the lower right segment being lit. If you go to the other side, there's the bottom segment being lit. There's the decimal point being lit, and so on. That's a seven-segment display, and it basically consists of LEDs, "light-emitting diodes." By the way, the sort of gray displays you see are liquid crystal displays. They involve polarization of light. More on those in the final lecture of this particular series.

Those are light-emitting diodes. Another application for light-emitting diodes: Here's an LED flashlight. It produces a very bright, very white light. White light LEDs are a fairly new invention, and they found their way into flashlights. This flashlight is both brighter, and lasts far longer than an incandescent flashlight, partly because LEDs have no filament to burn out.

More importantly, the battery lasts longer, because the LED is very efficient and converting electrical energy to visible light. An incandescent light bulb is very inefficient. It produces mostly heat. Sometime in the next decade or so, I can almost guarantee you that much better engineered LEDs, which are still in the development stage, will replace all incandescent light bulbs.

Cities have already saved enormous amounts of energy by replacing traffic lights, which used to consist of an ordinary light incandescent light bulb with a filter over it that wiped out most of the light, and made what was left come out red or green. It was horribly inefficient. Modern traffic lights in many states and cities are little tiny patterns—you can almost see the dots light patterns, the light of LEDs in appropriate colors—and are much more efficient. They save huge amounts of money on their electric bills. Those are LEDs.

The reverse can occur as well. In LED makes light, and you can also bring light into a PN junction, and make electricity. The other day, when I was outdoors, I showed you this photovoltaic panel. I will give you a little more sense, briefly now, of how that thing works. Let's take a deeper look at it.

The photovoltaic cell is basically just a PN junction. It's got that electric field in it, and it's not conducting. There's no electricity, no current flowing; but light comes in and the light energy can create an electron-hole pair. Now, that electron-whole pair finds itself at the junction with those strong electric fields, and an electric field will drive the hole downward and the electron upward. That makes a current flow. That is basically, ultimately the physics behind how a photovoltaic cell works. We all saw one in action the other day, actually running a little fan on a bright sunny day outside The Teaching Company here.

That's not what this lectures about, though. I want to end in the last few moments with a description of a device that is really important, and that's the transistor. Here's how we start with the transistor. There are many kinds of transistors: bipolar junction transistors, junction field-effect transistors, etc. This is called a *MOSFET*, a *metal-oxide-semiconductor-field effect transistor*. The name sounds complicated, but this is one of the simplest transistors to understand, which is why I'm using it here.

The MOSFET consists of a block of P-type material, with two little intrusions of N-type material. You'll see in the next lecture how you'd actually make this thing. It could equally well be N-type material with two intrusions of P-type material. That's the nice thing about semiconductors.

There's always this complementarity, and that sometimes serves as well, but I will just discuss this particular MOSFET.

There it is, and now I'm going to connect it to the outside world by hooking up a battery. Well, that was kind of a stupid thing to do, because no matter how I connect this battery, one of those junctions is forward-biased, and one of them is reverse-biased. Remember that if we connect the negative to the N, we've got a forward-biased junction. If we connect the positive to the N, we've got a reverse-biased junction.

There are two junctions in this transistor. Each of those little blocks of N-type material is connected to this P-type material, and there are junctions in between them. Consequently, no current will flow in this device. A forward junction on the left, a reverse-biased junction on the right—it would make no difference if I changed the battery around. All that would happen is that now I would have a reverse-biased junction on the left, and a forward-biased junction on the right. Still, the thing would conduct no electric current.

That's a pretty dumb device, but that's not all there is to a MOSFET transistor, or any transistor; because unlike a diode, the transistor has not just two places where you connect electricity to it—electric currents, electric voltages—but three. It's got a control electrode, and the key to a transistor— this is what makes a transistor so valuable—is that it allows one electric circuit to control another. That's the key element in computing, it's the key element in stereo amplifiers, and it's the key element in almost every electronic application—to have a device in which one circuit controls another.

How do we make that happen in the transistor? We added a thin layer of oxide. The wonderful thing about silicon is that if that P-type material is silicon, you simply expose it to oxygen and let it oxidize, it makes silicon dioxide sand, glass, quartz, which are an excellent insulators. You can build insulating layers on these materials, and we will talk more about how you do that in the next lecture.

Here we have this thin insulating layer, and then we coat a thin layer of metal on top of that. Now, let's see what happens if I take that thin layer of metal, and I connect it. We call that metal the *gate*, that thin insulating layer. If we take that gate—it's a gate because it's a control element—it decides whether electrons are going to get through this thing or not.

What happens if I take that gate and connect it to the positive terminal of the battery? Now, I would probably do more complicated things. There might be resistors, or I might go to a slightly different voltage, but the basic idea is this: What if I connect that gate to the positive terminal of the battery? Now, the gate gets positive charge on itself, so I've drawn some little pluses on the gate.

The nice thing about the MOSFET is that no electric current flows through the gate. So power, a product of voltage and current as we saw with a demonstration some lectures ago, there's no power at least being drawn by the gate circuit. It takes almost no power. Almost no current flows through that insulating layer of what is essentially glass or silicon dioxide.

However, the electrical attraction of the positive charge in that metal gate, for any negative charges that are sitting around in that P-type material. Now, we have to remember something. The P-type material is P-type because the majority charge carriers are holes, but there are still some *minority electrons*. They are there because of that random thermal excitation that produces electron-hole pairs.

There aren't a lot of them. They're spread throughout the whole P-type material, they're very diffuse, and they play very little role normally in the electrical conductivity of this material. Now—and again, this picture is not really to very good scale—I've got this thin metal layer that is positively charged. Those positive charges attract electrons in the P-type material, again, those few minority carried electrons, and it attracts into the region right under the gate—a semiconductor with lots of electrons in its N-type material. And if this device is engineered right, and you put in a voltage on it, you will pull enough electrons into that "channel" between the two little permanent intrusions that you make the whole channel, N.

Consequently, we have just a block of N-type conductors. We don't have any junctions anymore to block the flow of current, and current can flow through this device. That's how the transistor fundamentally works. You apply a voltage to the gate, you put positive charge on the gate, and you therefore make the channel conducting, and there you have a transistor that has been turned on.

Now, how well the transistor conducts depends on how much voltage you put on the gate. If you put a little bit, you will only attract a few electrons, and you won't get much conduction. If you put a higher voltage, you'll get lots of electrons. There are two basic modes to operate this transistor in, or any other electronic device. In the so-called *analog* mode, which is what is going to your stereo amplifier for example, you use a variation in the voltage at the gate to control a great big current that may be flowing through that channel.

Put a little bit of voltage, and a little bit of current flows. Put a big voltage, and a big current flows. An audio signal with varying voltages like we saw on the oscilloscope when we had our musicians playing instruments, and a varying current flows that follows that. That current could be strong enough to drive a loudspeaker. That's the essence of amplification. In amplification, you vary the gate voltage continuously to vary the current in that channel continuously.

In digital electronics, on the other hand, we want that dramatic difference between "one" and "zero," "yes" and "no," "on" and "off," and so we use the gate as a switch. If you put zero volts on the gate, no current flows, and the thing is off. If you put a certain substantial voltage on the gate, the transistor turns what we say is *fully on*, and it conducts. You therefore have just two states for the transistor, and it basically becomes an on-off switch. That's how we're going to be using transistors in digital electronics.

It's possible, by the way, to build, as I said, a complementary transistor, and in many cases, putting complimentary transistors together gives you a much more efficient semiconductor logic circuit, or a much more efficient amplifier. Chances are that the output in the final stages of your amplifier has complementary transistors driving the speaker in each direction.

Transistors are remarkable electronic control devices, and let me end by showing you a couple of transistors, and then going to our required summary slide. Here are some transistors. A little one, with again, three leads; a medium one—this one can handle a watt or so of power—and here's a big power transistor such as might swing a few tens of watts in the output of a power amplifier. It's mounted on a big heat sink, because it will get quite hot as it is operating.

Those are some individual transistors, and let's move now from atom to transistor. There's where we were at the end of the last lecture. We had gotten from atoms to PN junctions, to P and N materials. This lecture, we built PN junctions. We discovered that they worked as one-way valves, and from that, we were able to build that transistor.

Lecture Twenty-One
Building the Electronics Revolution

Scope: Transistors, invented in the 1950s, were a great improvement over their predecessors, vacuum tubes. But it was still tedious to connect individual transistors into a circuit that performed a useful function. The revolution that enabled modern electronics came in the early 1960s, when engineers learned to combine multiple transistors and other electronic devices on a single piece, or *chip*, of silicon. The first such *integrated circuits* contained only two transistors. By 1965, the number of transistors on an integrated circuit had increased to about 100, which prompted Gordon Moore to propose *Moore's law*, stating that the number of transistors per integrated circuit would grow exponentially, doubling each year or two. Moore's law has now held for more than four decades, and today's most advanced integrated circuits contain a billion transistors. Each transistor is smaller than an influenza virus! Not only are transistors getting smaller and more densely packed, but they're also getting faster at switching from the conducting to the nonconducting state, enabling ever-faster computers. That's why the new computer you bought last year is already obsolete!

Eventually, the continuing miniaturization of semiconductor electronics will run into fundamental limitations set by quantum physics. Some startling alternatives are already in the works.

Outline

I. Transistors aren't the first devices that use one electric circuit to control another.

   A. *Electromechanical relays* use an electromagnet to operate a mechanical switch. The current to the electromagnet thus controls the current through the switch. Relays are heavy, expensive, and slow.

   B. *Vacuum tubes*, invented in the early 20th century, control the flow of electrons in an evacuated glass tube. They contain a hot filament, similar to a lightbulb's, to provide a source of electrons. Tubes are large, bulky, and fragile; consume a lot of power; and burn out frequently.

1. One of the first electronic computers, ENIAC (Electronic Numerical Integrator and Computer, 1946), contained some 19,000 vacuum tubes and more than 1,000 relays. It filled an entire room and consumed electrical energy at the rate of 200,000 watts. Yet its computing power was far less than the smallest of today's personal computers.

2. Although vacuum tubes have been largely superseded by transistors, they are still used in some specialized applications, including high-power radio transmitters and microwave sources. Some audiophiles prefer tube amplifiers. And the "picture tube," a specialized vacuum tube, is still used for video display in television and computer monitors—although that is rapidly changing.

II. Transistors are a great improvement over vacuum tubes, being small, robust, long lasting, and low in power consumption. However, the electronics revolution is fueled not by individual transistors but by our ability to pack thousands, millions, and even billions of transistors on a single "chip" of silicon.

A. The first *integrated circuits*, developed in the late 1950s, contained only a few transistors, other electronic components, and the electrical conductors that joined them into a useful circuit. Although integrated-circuit design is now the province of engineers, physics is at the root of it: Jack S. Kilby of Texas Instruments shared the 2000 Nobel Prize in Physics for his role in the invention of the integrated circuit.

B. In 1965, Gordon Moore suggested that the number of transistors on a single integrated circuit would double roughly every year or two. *Moore's law* has held through four decades, and today's most advanced integrated circuits contain about a billion transistors. This is true exponential growth!

C. Linked with the increase in the number of transistors on a chip is an increase in the complexity of circuits that can be built and in the speed with which circuits carry out their functions—and a drop in price per transistor. That's why your new, more powerful computer doesn't cost any more than your old, obsolete one did.

D. Today, integrated circuits are ubiquitous. Specialized chips operate our cell phones, our children's electronic toys, our DVD players, the digital controls on our washing machines and stoves, our robotic vacuum cleaners, our thermostats and home energy control systems, our cars' fuel and pollution control systems. Chips implanted under our pets' skin identify lost dogs and cats. Heart pacemakers contain chips that can be reprogrammed as the patient's needs change. Increasingly, formerly mundane devices become "smart" with the addition of specialized integrated circuit chips. And, of course, our computers are full of chips.

III. Manufacture of integrated circuits involves some of the most technologically advanced equipment and procedures. Ultra-pure materials and the utmost cleanliness are essential for high yields of reliable circuits.

    A. Chip manufacture begins with the preparation of pure silicon crystals.

        1. Pure elemental silicon (produced from sand) is melted in a furnace. A carefully prepared "seed" crystal is lowered into the molten silicon, then withdrawn at a precise rate while being rotated. Silicon crystallizes on the seed and the growing crystal as it's removed from the melt. The result is a long cylindrical ingot of pure silicon.

        2. The ingots are sawed into thin wafers, up to about a foot in diameter. The wafers are polished to make them perfectly flat. Lots of waste here!

    B. A variety of processes are used to dope the silicon, making regions of P- and N-type silicon to form transistors, other electronic components, and their electrical interconnections. Most of this work is based in *photolithography*.

        1. A typical process starts with the growth of a thin layer of ultra-pure silicon on the wafer—even purer than the wafer crystal itself. (Remember from Lecture Nineteen how dramatically contaminants affect the electrical characteristics of the silicon.)

        2. The wafer is heated and exposed to oxygen. This adds a thin layer of silicon dioxide. The silicon dioxide layer is coated with a light-sensitive material called *photoresist*.

        3. A mask is prepared, typically with a laser beam or electron beam etching a metal coating off a plate of high-quality glass. The mask outlines features on what will be one of many layers in the finished chip.

4. Light is shined through the mask, exposing regions of the photoresist. As Moore's law shrinks the size of individual circuit elements, the wavelength of the light needed to transfer patterns to the silicon wafer shrinks, too. Today's high-tech chips feature transistors less than 1/10 of a micron across, much smaller than the wavelength of visible light. As a result, modern chip-making processes must use ultraviolet light, usually from excimer lasers similar to those used in laser vision correction. Typically, a wafer can hold several hundred identical integrated circuits; thus, the mask is stepped repeatedly over the wafer to produce multiple circuits.

5. When the photoresist is developed, it leaves a layer over those regions that were exposed to light. An acid is then used to etch away the oxide layer. Exposure to a gas or beam of dopant atoms then creates the desired P or N type of semiconductor in exposed regions. The process is repeated to make additional layers that complete transistors and interconnect them to form the integrated circuit.

6. Individual circuits are tested, then cut from the wafer, then mounted in sealed packages for insertion onto printed circuit boards.

IV. Eventually, the uncertainty principle of quantum physics will impose a lower limit on the size of conventional transistors. The future will probably require alternative technologies to perform the functions of today's semiconductor electronics.

A. Beyond that scale, *quantum computing*, which actually exploits ambiguity at the subatomic scale, may enable even greater advances in computing speed and power.

B. Researchers are also experimenting with "circuits" that use biological materials, such as DNA, to carry out computational operations. Tiny "DNA computers" might be placed inside biological cells for monitoring and control of cell functions. DNA computing has developed to the point where DNA-based computers can solve simple computational problems, thus raising the prospect of general-purpose computers based in molecular technology at much smaller scales than the current silicon technology.

V. Pause and consolidate: We're on a hierarchical journey from atom to computer. With each step, we "hide" the next lower level of complexity in a single symbol.

    A. Starting with the atomic structure of a silicon crystal, we developed P- and N-type semiconductors. They show up symbolically as reddish and bluish colored blocks, respectively, in diagrams of PN junctions and transistors.

    B. Now, we've seen how to put millions or even billions of transistors on a single chip of silicon; thus, a picture of a completed chip becomes our highest level symbol.

    C. What we haven't done is to connect the transistors together in a meaningful way. That's what comes next.

Suggested Reading:

Scientific American, fall 1997, special issue: "The Solid-State Century."

Going Deeper:

Altan Ferendeci, *Physical Foundations of Solid State and Electron Devices*.

Questions to Consider:

1. The maximum number of transistors on an integrated circuit went from about 1,000 in 1970 to about a billion in 2005. Given this growth, roughly what is the time needed to double the number of transistors?

2. At the start of the 21st century, the most advanced integrated-circuit manufacturing involved features—such as individual transistors— about 130 nanometers in size (1 nanometer is one-billionth of a meter). By mid-decade, that had been reduced to 90 nanometers. At roughly the same time, standard silicon wafers went from 8 inches in diameter to 12 inches in diameter. By what percentage did both changes together increase the number of integrated circuits that could fit on a wafer? Assume—incorrectly—no increase in circuit complexity.

Lecture Twenty-One—Transcript
Building the Electronics Revolution

Lecture Twenty-One, "Building the Electronics Revolution." Now we have the transistor; and in the previous lecture, we probably saw more detailed physics than anywhere else in this entire course. This lecture is going to be more or less the opposite. We're going to talk about manufacturing. We're going to talk, even, a little bit, about economics. We're going to talk about a whole industry, and how it all operates, and how it's come to permeate our whole world, but it's an industry that ultimately based in the physics of semiconductor electronics.

In the transistor, we have this single control element that can amplify a signal. We put in a small signal, and it changes slightly. A bigger signal changes in proportion, and that's how we get amplification; but more importantly, for our purposes in heading from atom to computer, we're going to talk about using transistors as switches that can be either on or off. And again, we use that voltage as a gate to decide whether that transistor is going to be allowed to conduct electricity by having its channel act as a kind of semiconductor, or whether it will block electric current. It's an on-off decision that we're going to be making.

Individual transistors are not enough. What has really propelled the electronics revolution is the ability to put lots of transistors automatically on the same single piece of silicon. I remember several stages in the process to getting there. Some of you may remember these historically as well. In the old days, one wired, by hand, individual pieces of individual electronic components, one to the other. One built them on complicated-looking, usually metal, chassis in which you had little pieces of insulating material with metal tags on them, and you could attach wires there, and you would solder them together. What a complicated job.

A little while later, we developed printed circuits. This was a flat piece of insulating material on which copper was coated, and then copper was etched off to leave a pattern of interconnections, and individual electronic devices, resistors, transistors, or whatever, were dropped into the holes, and soldered together. Printed circuit boards still exist, but now, they don't contain very many individual transistors and so on, as I will get to in a moment.

Some of you may date yourselves by having built electronic circuits yourself. There were a number of manufacturers who made electronic kits. I remember that when I was in college, I earned spare change from people who wanted new stereo systems. They could buy the stereo amplifier completely built, or for about half the price, they could buy it in kit form, and put together. They didn't want to put it together, though, so I would put it together, and I would charge half the difference between the two. Wiring electronic circuits was something a lot of people, particularly electronics hobbyists, used to do a lot.

The electronics revolution has been some ways, unfortunately, obviated the need to do that. Today's electronics are prewired, largely because we have managed to put thousands, hundreds of thousands, millions, hundreds of millions, and by the middle years of the first decade of the 21st century, billions of electronic components, particularly transistors, on a single chip of silicon. Today's lecture is about how we do that, and what some of the economics of that are that have powered this remarkable revolution in which we now have "smart" devices in our cameras, and refrigerators— well, not refrigerators, but we're about to have refrigerators that know how much milk is left, and they will tell you when to buy more milk. That's all powered by the ability to put thousands, millions, and increasingly, billions of transistors on a single chip of silicon.

Now, I emphasized last time that the role of the transistor is to be a control element. It lets one electronic circuit control another. Transistors are not the first such devices to do so, and I want to take you on a little historical tour of some of the other devices that we have had that have served the same function. And in principle, but no way in practice, we could build all of today's devices, today's electronics, out of these other things I'm describing.

Let me begin with a simple device that is still in use for certain applications. That's the so-called *electromechnical relay*. I have one here, and all this relay has is a coiled wire. You can kind of see this golden part. That's a coil. There's a coil of wire, and that coiled wire makes a little electromagnet. That coil of wire is in the center of a little flapper, which is just a switch that can either closer or open in an electrical circuit.

An electromechanical relay works by passing current through the coil. It becomes an electromagnetic, and that closes the switch. By the way, I describe such device earlier. That was basically how a circuit breaker works, only a circuit breaker controls its own circuit, where is this electromechanical relay is typically used to control another circuit.

There are many examples of relays. When you turn on the ignition switch in your car, that switch doesn't directly control the flow of 100 or more amperes of current from the battery to the starter motor. That switch couldn't handle that kind of current. It would overheat. You wouldn't want to run wires that long. So, what that switch does is to close a relay that ultimately connects the battery to the starter motor, which starts the car. Increasingly, other parts of cars—for instance, when you turn on the headlights, you're probably ultimately turning on a relay.

Electromechanical relays are still in use. They're cheap and rugged. They're not very sophisticated, and in particular, they aren't very fast. It takes a fraction of a second to close the switch on an electromechanical relay; but in principle, because an electromechanical relay is in on-off device, we could in principle build an entire computer from billions of these. We wouldn't want to, but we could.

Electromechanical relays were among the earliest of such devices. By the way, the early telegraph system—dot, dash, dot, dash—and the sounds you hear were basically operated by a device like a relay, again, a coiled wire that simply pulled down something else, some metal piece, and made that noise, the "dot-dash" kind of noise.

Around the turn of the 20th century, out of experiments, for example, that led to the discovery of the electron, which involved trying to pass electric currents between different electrodes in a vacuum by means of electrons that flowed through the vacuum from one electron to the other—out of that came the electronic vacuum tube. The vacuum tube dominated electronics for the first half of the 20th century or a little bit more, well into the '50s and '60s.

Vacuum tubes are still used in some applications, and I want to talk a little bit about vacuum tubes, give you some history of them, and give you a sense of how they work. A vacuum tube is actually easier to understand in some ways than the transistor. I have some examples of vacuum tubes here. An early vacuum tube, probably from around the 1920s, quite a large, ungainly-looking structure.

As vacuum tubes evolved, they got a little smaller. The final vacuum tubes before they went out of popularity, most of them looked like this very small one, but they all share a common future. They have a glass envelope. Air has been evacuated from the envelope, and there are metallic electrodes inside.

Vacuum tubes are kind of bulky. As we will see a minute, they also get hot. They're fragile. They burn out frequently, as we will see in a moment. They have to be replaced. Therefore, they have to plug into particular sockets. They aren't something that you leave in for the lifetime of the device, so they are large, they are bulky, they're fragile, and they consume a lot of power, but they work on a very simple principle, which I will just take a minute to show you.

Before we do, I want to give you a quote that's going to sort of start us on this history from vacuum tubes, to semiconductors, and ultimately, to putting lots of semiconductors on a single chip.

This is Richard Feynman, and he was speaking in 1959, at about the time vacuum tubes were going out, and transistors were coming in. However, transistors were still individual devices that had to be wired together. Feynman knew about computers, and he was talking about computing systems. He said, "I know that computing machines are very large. They fill rooms." Now, he's talking about some of the first computers that were built in the 1940s and 1950s.

One of the first was ENIAC, "Electronic Numerical Integrator and Computer." ENIAC had something like 19,000 vacuum tubes, more or less of this size. It had 1,000 electromechanical relays. It filled an entire room, as Feynman was saying, and consumed power at the rate of 200,000 watts. It was very slow to make calculations. It was always breaking down. Some of those 19,000 tubes were always bring out, so it was very unreliable, and it was far less powerful than this G4-based laptop computer I've got sitting here, which, although only two years old, is already not the current technology.

That's how bad computers were in the 1950s, and so Feynman says, "They're very big, and they fill rooms. Why can't we make them very small? Make them with little wires, little elements, and by little, I mean little. For example, a wire should only be 10 to 100 atoms in diameter, and the circuits should be a few thousand angstroms across."

Now, an angstrom is one over one, with 10 zeros, of a meter, and it's a tiny distance. An angstrom is about the size of an atom, so he says that the circuit should be about 1,000 angstroms across. One thousand angstroms is 100 nanometers, which is the more modern term used for very small-scale things, nanotechnology. Ironically today—or not ironically, but presciently—today's integrated circuits of the current technology in actual manufacturing, uses 90-nanometer technology. That's 10% smaller than what Feynman was suggesting, if you interpret his "a few thousand" to mean 1,000. We are quite a bit smaller than that if he really meant a few thousand.

"Everybody who has analyzed the logical theory of computers has come to the conclusion that the possibilities are very interesting, if they could be made to be more complicated by several orders of magnitude. Instead of 19,000 vacuum tubes, you could have 190,000 vacuum tubes, or 200,000, or two million."

Wow, that would be complicated. Well, now we're at one billion vacuum tube equivalents in our computer circuits, so we have gone way past Feynman. "If they had millions of times as many elements,"—well, let's see, a million times 10,000 is 10 billion. We are getting there. We are not quite there yet –"then, they could make judgments." Already, some of our computer chips are as complicated as the brains of simple insects."

Feynman, then, was very prescient; he didn't quite see how to do it, but he thought we ought to be getting there. Vacuum tubes have evolved into the 1950s and '60s into these relatively small vacuum tubes, but then the transistor basically superseded them. There are a few cases where vacuum tubes are still used. I have mentioned the magnetron before, a special vacuum tube that generates the microwaves in the microwave oven. Some high-powered radio transmitters still use vacuum tubes, and there are audiophiles who claim—and I do not wish to get into this controversy, because I'm a bit of a skeptic—that there is something more mellow about the sound of audio produced by vacuum tubes. Now, in my purest sense, I think if you're going to have an audio amplifier, it ought to reproduce faithfully whatever signal you put into it, but given that that doesn't happen perfectly, maybe there's some truth to the fact that vacuum tubes make things sound a little better. I don't want to go there, though. There are still people who use vacuum tubes, though.

Another place. Vacuum tubes are still in use is in traditional televisions and computer monitors with a so-called *cathode-ray tubes*, or picture tubes. It's a special vacuum tube that accelerates electrons. They bombard a phosphorescent screen at the front of the tube, and that's what makes the picture you see. Those, as I've indicated before though, are rapidly on the way out and are being replaced by LCD, plasma displays, and other devices.

In the previous lecture, I introduced transistors, and how they work. They are an enormous improvement over vacuum tubes, but they work by a more complicated physics, and I want to take a quick look at the physics of the vacuum tube, and then we will move on to transistors.

Here's a vacuum tube, a typical vacuum tube, very simple vacuum tube. That circle represents the closed envelope that I talked about. At the bottom, that sort of triangular looking thing is a filament, and you run electric current through the filament like the filament of a light bulb, and you heat it up. Electrons boil off, and then, if you connect a battery so that the positive end is at the uppermost electrode in the vacuum tube so that becomes positive, you connect the other end so it becomes negative. Electrons boiling off the filament will be attracted through the tube, they will hit the plate, and electric current will flow. That's how a vacuum tube works. By the way, that alone makes the vacuum tube a diode that serves the same function as our semiconductor diode, which passes current in one direction, but not another.

However, here's a third electrode in the middle. It's called a grid, and I've indicated it by a dashed line, because it's actually a grid work that lets electrons go through it. If I place a negative charge on that grid, then with a battery with a negative terminal connected there, the grid will repel electrons coming off of the filament, and they won't be allowed to get to the plate. The thing will then block the flow of electric current. I can therefore turn the current on or off with this grid, just as I can turn the transistor's current on or off with the gate, or I can adjust the current by adjusting the voltage on the grid.

That's how a vacuum tube works. It's basically like a transistor. It's a device that basically controls the flow of one circuit with another, but it is big and bulky. That filament burns out, and consumes a lot of power. It gets very hot, needs to be cool. It burns out, it's fragile, it's expensive, it's complicated to build, it has to be almost wired in by hand, it plugs into a socket, and it's a complicated device.

I don't want to go there. We have transistors now instead, and we've had transistors since the 1950s and '60s. Here are a host of transistors. I showed you them in a previous lecture. We know how they work, and they are tiny compared to vacuum tubes. Look at the smallest vacuum tube alongside a typical transistor. The transistor probably has 100th the volume of the vacuum tube, and where the vacuum tube has a lot of pins coming out of the bottom because it needs connections to its filament, and so on, the transistor has only three: One for the control, and two for the controlled. So, these two things, the vacuum tube and the transistor, are conceptually similar, but practically, they are very different.

Still, if I were to build a circuit out of either vacuum tubes or transistors, I'd have to connect a lot of transistors together, and the key to the electronics revolution is our ability to connect lots of transistors together on a single chip of silicon, so that we don't actually wire a circuit anymore. We build a chip, which, as I said, contains many, many transistors.

The design of *integrated circuits*, as these things are called, is clearly the province of engineers. This is not physics anymore, although as we will see, there are some physics principles that we'll use when we get to the limits of how small we can make these things. I should point out though that it is also the province of physics. Jack Kilby, of Texas Instruments, for example, shared the 2000 Nobel Prize in physics for his contributions in the late 1950s to the development of integrated circuits.

I have here are a few integrated circuits. Here's a simple logic circuit. This is actually still in use. Due to the size of the device, this was probably available first in the mid to late 1970s. It probably contains several dozen transistors.

Here's a bigger integrated circuit from about the same era. It probably contains a few thousand transistors. These are actually tiny circuit boards—two of them—with several different integrated circuits mounted on each. They are actually memory modules. The one on the left us from an earlier era, and each one has about a few hundred thousand transistors. The one on the right is from a later era, and each integrated circuit on there has a few tens of millions of transistors in it.

Finally, I have a complete *motherboard*, as it's called, the main processing board out of a personal computer. You can see the Pentium chip here. This is not the most current version, but this one probably has tens to hundreds of millions of transistors in the chip.

Each of those transistors serves the same function as one of these vacuum tubes does. With this, we have realized Feynman's dream. We have realized the complexity that he thought we ought to get to, in order to make computers really special.

Now, in 1965, an engineer named Gordon Moore noticed even that early that integrated circuits had been getting better, and better, and better, and he proposed that the engineering of integrated circuits was going to improve to the point that every roughly one to two years—and it's about every 18 months, as it turns out—the number of transistors that we can fit on a single integrated circuit would roughly double.

Here's a graph of Moore's law, from about 1970, on to about 2005. The individual triangles are individual actual chips that were manufactured and put into production. The line is sort of a trend. When you see a straight-line on a logarithmic graph, and you look at the left hand axis of this graph, it goes from 1,000 at the bottom, the next tick mark is 10,000, next is 100,000, then one million, 10 million, 100 million, one billion.

Each tick mark goes up by a factor of 10. That's a logarithmic scale. When you see a straight-line on a logarithmic scale—that implies exponential growth. So, the number of transistors that we can fit onto a chip is exponentially doubling about once every 1.5 years. That, by the way, is why your new, powerful computer that you just bought is obsolete almost as soon as you buy it. Next year, or next month, there's going to be one that's twice as powerful—or much more powerful anyway—and it's going to cost either the same or a little bit less. This is because even as the number of transistors we can fit on a chip increases, the cost for each transistor drops lower, and lower, and lower. They started out at over $0.10 to $1.00 per transistor, and by 2005, will be down to about one millionth of one cent, which is 10 billionths of one dollar, 10 nano-dollars, if you want like, in today's "nano" language.

The cost per transistor has therefore been dropping dramatically, and integrated circuits today are ubiquitous. There are specialized integrated circuits in your cell phone, specialized integrated circuits that are the receiver circuit of your FM radio, your DVD players have specialized integrated circuits that decode the stream of bits coming off of the DVD, and other ones that do the business I mentioned before, of focusing lens, and keeping it—all customized integrated circuits. Our washing machines, our robotic vacuum cleaners, our thermostats, and our home energy control systems—all of these things now are smart because they contain specialized integrated circuits—your heart pacemaker, your pet may have a chip in it that tells you if your pet has gotten lost. Soon, products in the store will have integrated circuits that will emit radio signals. It will tell you what that product is and it will also serve security needs.

This is wonderful stuff, and we're all involved with it, because we all use computers, cell phones, televisions, and all these other things that now have increasingly smarter and more complex integrated circuits in them. How do we build these integrated circuits? Well, that's one of the most high-tech things we do. Here's an example of a clean factory. I have to emphasize, "clean," because, remember that silicon has to become so that we can dope it exactly right at the level of one doping atom in 10 million silicon atoms.

We have to keep things really pure, so these semiconductors are manufactured in these wonderfully clean facilities. Here, you see these workers dressed up in these Moon suit-like things to keep from contaminating anything with stuff off of their own bodies. It's a remarkable process, and the process begins with the preparation of pure silicon crystals. We get pure silicon ultimately derived from sand—again, that very commonplace material—so we make pure elemental silicon.

We melt down metallized silicon, and then in a process that involves rotating and putting a seed crystal down on a block, dropping on a rod, and dropping it into the molten silicon, we slowly draw out the molten silicon, all the while rotating it, cooling it at just the right rate, so that we get a cylindrical ingot of pure, crystalline silicon. It has to be pure, and it has to be crystalline, and most applications, it has to be a single crystal. Even though silicon is the second most abundant element on Earth, pure, semiconductor grade silicon is very expensive, simply because of the process of making this pure silicon: it has to be really pure, because we know what contaminants do—later, we want to contaminate it, but we don't want it contaminated to start with—and the fact that it has to be crystalline, all atoms have to be sharing the same common crystal arrangement. If we have gaps in the crystal structure, we have problems with the electronics of the semiconductors.

What happens if we get one of these ingots of pure silicon out? Well, we slice it up with the specialized saws into these thin wafers. An awful lot of it is lost in that sawing process. People are also working very hard to develop ways of growing thin-film silicon. In some cases, we've succeeded in doing that for some applications. For example, in my solar panel over here, as I mentioned to you when I showed it to you outdoors is, in fact, polycrystalline silicon, and you can see that because of the mottled appearance of it, that reduces the efficiency of the photovoltaic device, but makes it far cheaper to build. For semiconductor electronics though, we slice this silicon ingot into these very thin wafers, about a millimeter thick, and we then process those wafers. I'm going to show you briefly the process we use to go from there to making individual transistors.

It is a process called *photolithography*. Here's a wafer. Typically, these wafers are 8, 10, or 12 inches in diameter. The size has been growing as the semiconductor revolution has advanced. In my own state of Vermont, we had probably what was the world's state-of-the-art facility for making eight-inch semiconductor wafers, that IBM facility there has now has been superseded by a facility in New York that makes larger wafers. We're trying to get more of these chips on a wafer.

This picture shows you a grid-like pattern of rectangles. Those are to be individual integrated circuit chips, each of which will be identical to each other one, and will serve a certain function. They might be a computer processor, they might be a memory chip, or whatever, and typically, there are more like 100 of them on a single wafer, not just a few, as I've shown you. I guess I've shown you 16. What happens? We've got this wafer. We expose it to oxygen, and then grow a thin layer of silicon dioxide on it. That silicon dioxide layer is coated with a photo resistive material that responds to light, and is like camera film, and can be developed when you shine light on it.

Then, we develop a mask. Actually, there are several masks in the complicated manufacturing process, but the masks reflect what we want to put on this ultimately silicon substrate. We will actually grow some more pure silicon on first, even purer than the silicon that came out of the oven.

Here's the mask, and we're going to shine light through the mask. I've marked is as UV light from a UV laser, although there may be other sources. It used to be that this could be visible light, but as I mentioned in the first module, you can't image, make shadows, or anything else like that with light. You can't make objects that are comparable to, or smaller than, the wavelength of light, and by now, we're working with 90 nanometer basic device sizes. Basic transistors are only 90 nanometers across. The wavelength of visible light, as I showed you in the last lecture, and in the previous module, is somewhere between 700 and 300 or so nanometers, four or seven, to eight or ten times as big as the devices we are now making.

We can't use visible light anymore. We have to use ultraviolet light, which has a much shorter wavelength. So, that diffraction limit that I talked about in the first module is coming to this again, here. As we continue to make these smaller and smaller, we have to use higher and higher frequency, shorter and shorter wavelength light. We expose the mask, and that puts onto this wafer, a pattern. The mask is typically much bigger than the ultimate size of the integrated circuit, so the focusing process of the light reduces the size to what we actually want on the chip. When the *photoresist* is developed, there's a layer over those regions where it's exposed to light, and then, we begin the process of a fabricating the individual parts that make the layers of all the many individual transistors. I'm just going to give you the briefest hint of how this happens, but you can imagine how the whole process must go on over, and over, and over again, as we build up a more complicated structure on what is ultimately this wafer of silicon.

Here's how we might go the transistor. We want to build one of those MOSFET transistors like I described in the last lecture. Here's a slab of material that we have already exposed to something that has made it P-type. Already, that silicon has been exposed to, perhaps, an arsine gas, which contains arsenic, and arsenic has diffused into there. That's what gives us the P-type material.

Now, we have this oxide layer, and the photoresist, and we've put the mask on. We've developed the photoresist, and you see those two gaps, where the photoresist is gone. The photoresist is resistant. It is not only photosensitive to light, but it resists acids, so now, we etch away, typically with acids or other means, where the photoresist is not there. That takes away that oxide layer down to the P-type material underneath.

Then, by means again of a variety of techniques, increasingly high-energy beams of material, but more commonly, simply exposure to a gas that might contain the right material—here's a gas containing the dopant, this is a vapor, or perhaps it's a beam of dopant atoms. The material is exposed; where the oxide layer is present—the photoresist has now been removed, but where that oxide layer is present the material—this new material is not allowed to get into the semiconductor substrate, the P-type material, but through those two gaps, it is. Diffusion again occurs, and we have a high concentration of the dopant material right at the surface of the P-type block, and so, that dopant, which is an N-type material, diffuses. Did I say arsenic before for the P-type? I think I meant the other way around. This might have been arsine gas, and the arsenic diffuses to make these N-type regions. By controlling the concentration of the gas and how long we expose it, we control how deep in those N-type blocks go, and the level of concentration. There we are, with those exquisite controls on those electrical properties of these materials.

Then what? Well, we play again with some type of photo resistant material, so it only happens here—that metal layer on which is going to become the gate. Now, we almost have the complete transistor that I described in the previous lecture, but we have to make contact to it, so we could put some more metallic layers on that make the contacts to the N-type material. Then, we probably could put some more oxide layers on that to serve as insulators, which separate the now metallic contact to the gate.

While this is going on, another 999 million, or however many other transistors there are in this chip, are all having the same thing happen at the same time. In addition, these things that show up here as gold or yellow are the interconnections. They are being made to interconnect the individual transistors in ways I'll describe in the next lecture, to make the final circuit that does exactly what we want it to do.

Eventually, we reach the complete wafer. It looks like that. It's got this pattern of rectangles on it. That wafer might be eight inches across, and each of these little rectangles is one of these chips. The chips are cut apart. There's a microphotograph of what is on an individual chip, those patterns on there. You don't see individual transistors at this scale, but you do see a kind of patterning, which represents the way transistors are arranged. The chip is ultimately packaged in a covering package to protect it, and the external wires and connections to the external circuitry are ultimately connected, and there we go. Ultimately, we have this single chip.

Before I end up with my final slide that shows how we got from atoms here, let me just say a few other things. Although Moore's law has held now for a number of decades, from the mid-1960s to now, ultimately, we will run into quantum mechanical uncertainty principle limitations on how small we can make these things. We aren't quite there yet, but will probably be there in a matter of a decade or so. People are also experimenting with circuits made with DNA, for example. Already, it has been demonstrated that DNA can do actual computations, solve simple constitutional problems, so we might ultimately get molecular scale computers that are based on principles that govern life.

There are a number of processes of superconducting computers that involve something called *Josephson junctions*, and quantum tunneling, and there are quantum computers that I'm not even going to go into that have crazy logic, due to the multiple states of quantum physics. There will, then, eventually be some replacement for this silicon-based technology, but we aren't there yet. We are still in the silicon, and we will be there for a number of years to come, and probably a decade or so.

Let's pause and consolidate here. We are on this hierarchical journey from atom all the way to computer, and we are now about halfway there, in some sense. We started with the structure of the silicon crystal, and the atom and the silicon crystal. We built up to P- and N-type semiconductors, and now, we've got these things formed into transistors on these chips. Our whole picture looks like this. Here's where we were at the end of the previous lecture. Silicon atom, silicon crystal. We made P- and N-types; we made P- and N-junctions, and transistors out of that. Here, we have learned how to put lots of transistors onto a single chip. There's the chip, and that chip can do useful things. How does that work? That's the topic of the next lecture.

Lecture Twenty-Two
Circuits—So Logical!

Scope: It takes just a few transistors to make circuits that perform simple logic functions. Such circuits are the building blocks of all digital electronics, including computers. Digital circuits store and process information in the form of binary numbers—sequences of ones and zeros in which each *bit* conveys a single piece of "either/or" information: on or off, one or zero, yes or no, high or low, true or false… The fundamental building blocks of computers are circuits that operate on such binary information, giving answers to such questions as "Are these two bits both 1?" or "Is either of these bits a 1?" and performing simple operations, such as changing a 1 to a 0 and vice versa. From multiple applications of such basic operations comes all the complex information processing that we expect from computers.

Outline

I.   There's a hierarchy of ways to know and use computers.

   A.   Everyday users are often unaware that they're using computers. Driving a car, where computers control the fuel system and the antilock brakes, is one example, as is using an automatic teller machine or a digital camera.

   B.   Most of us use "canned" computer programs, such as word processing programs, spreadsheets, and email programs, to make our computers do useful tasks.

   C.   Programming languages allow computer programmers to write the sequences of instructions that constitute useful programs. Programming languages range from very high level languages, in which a single command results in a powerful sequence of computer actions, to low-level languages that correspond to the limited number of simple, basic instructions that are "hard wired" into the computer's circuits. In the middle of this range are such languages as JAVA, C, BASIC, and FORTRAN, used by programmers to create higher level programs and by scientists to create everything from spacecraft guidance programs to global climate models. Many "canned" programs have some programming capability, as well.

II. At the most basic level, all that computers do is simple manipulations of binary numbers, which may represent actual numbers or text or other symbolic information. In the 32-bit personal computers of the late 1990s and early 2000s, those numbers contain 32 bits (binary digits). Newer computers use 64-bit numbers. More complex tasks—searching for a word in a document, removing "red eye" in a digital photo, calculating a spreadsheet, checking an email for spam or virus—are ultimately combinations of simpler tasks. Here, we'll start with the basics and build up to see how a computer can test two binary numbers to determine if they're equal. A bit of glossary: A group of eight binary digits is called a *byte*.

A. All digital information processing boils down to three simple logical functions on the binary digits 1 and 0 (which, again, can be interpreted to mean yes or no, true or false, and so on). These functions take one or two input values (either 1 or 0) and give a single output value (either 1 or 0).

1. The *AND* operation gives an output of 1 only if both its inputs are 1. Call the inputs A and B; then, the output is 1 only if A AND B are both 1. The operation can be described logically by its *truth table*. Any circuit or device that performs this logic function is called an *AND gate*. Regardless of how they're physically constructed, all AND gates are represented by a common symbol. A crude way to implement an AND gate is with hand-operated switches, a lightbulb, and a battery. The switches are connected one after the other (in series). If a closed switch and a lit bulb stand for logical 1, then only when both switches are closed does the lamp light.

2. An *OR gate* gives an output of 1 if either input is 1—that is, if input A OR B is 1. Putting the switches in parallel turns our lightbulb circuit into an OR gate.

3. The *NOT*, or *invert*, operation turns a 1 into a 0 and vice versa. Implementing a lightbulb NOT gate is also easy, although we need to add a *resistor*—just an electrical conductor that isn't perfect—to keep too much current from flowing through the switch when it's closed.

4. In practice, it's more convenient to make and to work with *NAND* (NOT AND) and *NOR* (NOT OR) gates, whose outputs are opposite those of the AND and OR gates. Connecting the two inputs of either gate together gives a NOT gate, which means that we really only need NAND and NOR to build all possible logic circuits. (In fact, we really need only one of these types.)

   5. Implementing NAND and NOR with transistors is easy. We'll dispense with the lightbulb and take the voltage at a battery's positive terminal to signify 1 and the voltage at the negative terminal to signify 0. In practice, computer circuits are almost this simple, but they use a few more transistors and avoid the resistor, that way consuming much less power.

   6. Equipped with NAND and NOR gates (and NOT gates, which can be made from either of the others), we have essentially everything that's needed to build a computer. From now on, we'll stop thinking about individual transistors and just work with logic symbols.

B. One of the many simple instructions a computer can execute is to compare two binary numbers to see whether or not they are equal. These might represent actual numbers, or they might represent text, coded as ASCII, where each character is an 8-bit byte. A search for a particular word in a text document, for example, tests each word in the text using a character-by-character comparison that is ultimately a test for equality of two binary numbers. We want a circuit that takes two binary numbers as its input and produces an output whose value is 1 only if the inputs are equal. We'll build gradually to that circuit.

   1. With our basic gates, we can make the *exclusive OR* function, whose output is 1 if either of its inputs is 1 but not both. Here, we use four NAND gates. We can confirm that this works by constructing the truth table. This circuit also forms the heart of a binary adder, as well as of our equality tester. It's so useful that we'll give the circuit its own symbol.

   2. Combining the outputs of several XOR gates through a NOR gate gives us our equality tester.

   3. We'd like to test pairs of binary numbers with more than two bits each. That's easy; we add an XOR gate for each additional bit and feed the outputs into a multiple-input NOR gate—made from transistors just like our two-input NOR gate but with additional transistors connected in parallel. The video shows one that compares two 8-bit numbers, which could represent two alphabetic characters in ASCII code.

C. The equality tester is one of many circuits that execute simple instructions inside a computer. All are built from the basic logic gates, themselves built from transistors, which are made from P- and N-type semiconductors, which are ultimately made of atoms arranged into crystal structures with carefully engineered amounts of impurities added.

Suggested Reading:

Marshall Brain, *Marshall Brain's How Stuff Works*, pp. 112 –113 ("How Bytes and Bits Work").

Going Deeper:

Paul Horowitz and Winfield Hill, *The Art of Electronics*, chapter 8 through p. 500.

Neil Storey, *Electronics: A Systems Approach*, chapter 9.

Questions to Consider:

1. Design an exclusive OR circuit using only NOR gates instead of NAND gates (this can be done with just four gates).

2. Using logic gates with two inputs each, design a four-input AND gate—that is, a circuit that has four inputs and produces an output value that is 1 only if all four inputs are 1.

3. Show that you can make a circuit that provides the NOR function using only NAND gates and NOT gates (which you know can be made using NAND). Thus, you really need only one kind of gate to make all logic circuits.

Lecture Twenty-Two—Transcript
Circuits—So Logical!

Lecture Twenty-Two: "Circuits—So Logical!" Now we know how to put the billion transistors onto a single chip of silicon, and we understand how transistors work, but how do we connect all of those billions of transistors so that they do the kinds of things we want them to do? Isn't that a hopeless task to think about?

Well, it turns out not to be a hopeless task, and the reason is that fundamentally, computers don't do anything very complicated. They perform very, very, very simple operations on strings of binary digits, zeros and ones, that might represent numbers, or might represent symbolic information like text. The operations they perform are so simple that we can understand them very easily, and see how to build them up into more complicated operations.

Today's lecture is going to take you from an introduction, to the very simplest of those operations that are performed on binary digits, binary numbers, to a fairly sophisticated circuit that will test two binary numbers of some length to see whether in fact they are equal or not, the kind of thing the computer would do in an everyday way.

Now, you may or may not be a computer user. You probably are, but even if you aren't a computer user there are many situations in which you, in fact, use computers, and don't know it. Slam on the brakes in your car, and the anti-lock brake system comes on. As I've indicated before, that's a computer that uses information from sensors at the wheels to decide which wheels are slipping, and which wheels are truly rolling, and keeps them all rolling so that you have the maximum amount of friction. The friction is provided by static friction, which occurs when a wheel is rolling without slipping.

Your car's fuel system is also computerized these days. There are sensors that measure how much carbon monoxide and how much oxygen are in the exhaust gases. Sensors determine exactly when the spark is firing. Sensors tell how much load the engine is under, and all of these things are used to adjust through a computer that calculates in almost instantaneous real-time what needs to be happening with that engine.

If you go to an automatic teller machine, the automatic teller machine is basically a computer. It has a very small keyboard, and only the few instructions it can do, but it spews out your $100, and that's what you want. A computer has made it do that.

Your thermostat, increasingly, is a computer. A digital camera has all kinds of computing in it, including computing that takes a very complex, large, memory-hogging image and reduces it very quickly to a much smaller size so that you can store lots of images on your camera. More about that in the next lecture, when we talk about computer memory.

The point is that you are using computers all the time, whether you know it or not. All of these computers ultimately are digital computers, and they operate by the same basic principles I'm going to describe here. That is, they perform very simple, logical operations on binary numbers.

Most of us, in fact, do use computers, and we tend to use canned computer programs—word processing programs, spreadsheet programs, drawing programs, photo-modifying programs, that modify our digital photographs, email programs that receive and send email, a whole host of programs—that we use for the individual tasks we like to do with computers.

These are canned programs. Somebody has written a program that do all of these things for you in simple ways so that you simply click on "spellcheck" and it checks the spelling, or maybe it checks the spelling automatically. It's doing a lot of these simple operations, but you don't see that going on. You only see the very high-level operations. Check and see if the word "dog" appears somewhere in this document. Change the color of this apple from green to red. Remove that redeye this picture, because I had the flash on the wrong setting, or something. Those are large-scale, complicated operations that break down into many, many of these simple operations, and all of us who use computers are executing those simple instructions, and doing those operations all the time.

There are other levels at which one can get into computers and tell them what to do and program them. Some people—there aren't very many left these days—actually program computers in the fundamental language that the computer itself knows. A real computer comes with a relatively small set of things that it really knows how to do: Add two binary numbers, multiply two numbers, move a binary number so that all of its digits shift one place to the right, compare two binary numbers and see if they are equal or not. Very simple things like that. That last one was actually developed in the course of this lecture.

There are some people who actually write programming in what is called *machine language*, this fundamental language that the computer program knows. Each instruction in a machine language program corresponds to an actual electronic operation, or a logical operation that the computers actually do.

Then, there are intermediate languages such as those that my colleagues and I frequently use, languages like Java, C, BASIC, FORTRAN. FORTRAN is that venerable but still used language for scientific programming. These languages have powerful statements that combine a number of these basic computerized operations into one statement. Raise a number to a power, for example. Well, that's a pretty complicated thing to do, and it takes a number of these fundamental computer steps.

There are also programs like Mathematica, or Matlab. Programs like that that work at a higher level, in which a single statement may do hundreds or thousands of computations very effectively.

There are a whole range of ways in which we use computers, and are using computers without even knowing we are using computers, through what most of us do, which is to use word processing programs, drawing programs, photo programs, that kind of thing; email programs, down through these more advanced computer languages like Mathematica, Matlab, things like that, down through the basic programming languages like BASIC, FORTRAN, C, Java, and then onto machine language. We actually know what that particular computer can do, and you are programming in those individual languages.

I wanted to give you that overview, because what I am about to introduce are these basic fundamental operations that the computer can do. In any computer software, from the most simple machine language program, up to the most complicated piece of expensive software you buy for your personal computer, they're all fundamentally built up from these fundamental logical building blocks.

In the same way that in the previous lecture that we built complicated integrated circuits from individual transistors, here we're going to build complicated operations on binary numbers, from these simple, simple logical operations that computers fundamentally do.

Now, I have mentioned that on the most fundamental level, all any computer ever does are simple operations of binary numbers. Typically, the more advanced of today's computers operate on 64 bits, a string of 64 ones and zeros, making one long binary number at once. Computers of the late 1990s and early 2000s, typically operated on 32 bits of information, and some of those computers are still around.

When some of the earliest personal computers came out in the 1980s, they operated on eight bits at once, so as computers get more advanced, and as we can store more information on an individual chip, more transistors on a chip, there's Moore's law in action. We can process bigger, and bigger, and bigger strings of binary information at once. These complicated tasks we are going to build up are ultimately combinations of these simple tasks acting on these so-called *words*, these 64 or 32-bit strings of binary numbers. I want to give you an example of a couple of ways in which we might interpret binary information.

First of all, binary information can be interpreted as simply numbers representing numerical information, and computers that are doing so-called *number-crunching*—calculating climate models, analyzing astronomical data, building an economic forecast, or whatever—are basically processing binary digits as numbers.

For instance, here's the binary number 1001. It's a 4-bit binary number. It has four digits. What is it equal to? Well, we want to convert it to decimal, and what we recognize is that just as in a decimal number where we have the ones place, the tens place, the hundreds place, the thousands place, in binary numbers, we have the ones place, the twos place, the two squared place, which corresponds to the hundreds place—two square is four, instead of 10 squared is 100. Then, there's the twos cubed place—10 cubed would be 1,000. Two cubed is eight, so that is the eights place. The way we interpret this binary number is to look from the right—or the left; it doesn't really matter. We have one times eight, zero times four, zero times two, and one times one. That's one times eight plus one times one. That's the number nine in decimal, but it is number 1, 0, 0, 1 in binary, and those are both perfectly equivalent and legitimate expressions of the same number. I'll have more to say about this in a subsequent lecture.

We can equally well represent text with this number. In particular, there's a code that is agreed-upon, an entered code that is used for representing text. It is the American Standard Code for Information Exchange. You may have something as complicated as a word processing program, usually when you save the file, it is saved in the format of that program, but it could also be saved as so-called ASCII text, which any computer can understand. It leaves off italicizing, and all these complicated formatting features, and just saves the raw text in this code.

For example, here's a string of three 8-bit binary numbers. By the way, an 8-bit binary number is called a *byte*. A bit and a byte. It is like you've taken a bite of this number, because that's a convenient unit for talking about processing, so 8-bit bytes. ASCII is really a 7-bit code. We tend to make it 8-bit these days. So more letters simply have a zero in front.

Each of these numbers, which we could convert in the same way that we did the decimal numbers, stand for a particular letter. That combination—00100100—whatever that turned out to be in decimal, stands for "D." If you looked up the ASCII code, you might see a listing that gave the decimal equivalent for D it might also give the binary equivalent for D, and it could also give the equivalent in a couple of other number schemes, which I'll get to in a subsequent lecture. 01001111 happens to correspond to "O." The last one, 01000111, is "G," so this is the word "dog," spelled out in ASCII code. It takes a total of 24 bits—three times eight—to spell out that three-character word.

Your word processing document, your Ph.D. thesis, your letter home to Mom, your description of your trip to the Grand Canyon that you are writing about in an email—that's ultimately what it is, a huge, long binary number where each string of eight bits represents a letter, blank, comma, or some other character that you use to write text. Whether it's letters, words, or numbers, all computer-processing boils down to simple logical operations, based on, again, these strings of sequences of binary digits interpreted either as letters or as numbers. Let's see how that works.

Ultimately, again, remember that what we have are ones and zeros. We can think of the ones as saying "true," "on," "yes" —interpreted any one of those ways. The zero is "off," "no," "false" —different ways of thinking about it. We have to be able to distinguish those two. If our electronic circuits can do that, we can process these things.

There are only a handful of basic logic operations that we need to do everything. In fact, there's only one that we need to do everything, and we will get to that in a little bit. I want to introduce three pretty easy logic operations to you though, and talk about how we build them from transistors, so that eventually, I would link it to the material we've gone through before.

Let's look at the first operation. The first operation is called logical *AND*. Any logical operation is easily described by a truth table. In the case of logical AND, and one of the other operations I'm about to introduce, the conceptually logical structure of this thing is that there are two inputs. I will call them A and B. And, there's an output. Here's the question. For logical AND, the question is, are both inputs true? Are both inputs one? Are both inputs yes? Are both inputs at a high voltage? "High voltage" means one. If they both are, then the output is also high. If A is high, and B is high, then output is high. If A is one, and B is one, the output is one. If they aren't both one, then the output is zero.

I'm going to build up the truth table for logical AND. We look at all the possible input combinations. Zero and A, zero and B. Are they both ones? No, so the output is zero. Zero on A, one on B. Are they both ones? ARE A one and B one? That is why it is called the AND operation. No. The output is zero. Look at one and zero. Obviously, they aren't both ones, so the output is zero. How about 11? Well, yes. A is one, and B is one, so the output, in that case, is one. It's logical AND.

We ascribe a symbol to the logical AND operation, and this symbol that I put on the right hand side here is not something that represents a particular electrical, electronic, or mechanical circuit. It simply represents this logical operation however you might care to implement it, mechanically or electronically, or whatever. That represents the logic operation. It shows two inputs ended output, and that kind of elongated D-shaped thing kind of tells you that we are talking about the AND operation.

How would you actually make the AND operation? Well, I have here a model that shows a very simple way to do it, and very soon, I will show you how to do with transistors. We'll keep this model in mind.

I have here a light bulb. The light bulb being "on" represents one, and the light bulb being off represents zero. I also have two switches, and there's simply a little metal flapper that is not connected right now, connecting these two metal points. When I swing it the other way, it makes an electrical connection, and the switch is closed. These two switches are connected in so-called "series." One comes right after the other, so current that flows through one switch has to flow through the next one to get to the light bulb.

I've had this made completely transparent, so that it's totally obvious what's going on. I'm going to put a battery in here, and I'm going to hook up the battery to this circuit. I'm going to close one of them—both switches are off now. They're open. They are zero. That represents 00. When the light is off, zero. I'm going to close one switch, and nothing happens. The electricity can't come through, because the other switch is open. If I close the other switch, of course, nothing happens. The output is still zero. This is the 1 0 state. That switch is closed, and is a one. That switch is open, and it's a zero. No good.

If I close them both, the light bulb lights. Those switches and this light bulb are implementing the AND function in a simple electric circuit with hand operated switches. If I look at the light bulbs, and decided what other switches were going to do, and flipped them all with my hands, I could in principle build up an enormous human operated computer. That's the logical AND operation.

I'm going to show you a couple more logic operations. Here's logic *OR*. It looks the same—I'll hook the battery up—except that the switches are in parallel now. That is, electricity can go through this switch, or this switch. Right now, both switches are open. Both are zeros. This is the 00 state. The OR function says that if either A or B is a one, make the outcome a one.

Neither A nor B is a one right now, so the output is zero; but, if I make B a one, the light lights. If I make A a one by closing that switch, the light lights. If I close both, the light also lights. This is the OR in the sense that if either one or both of them are closed, if either input or both are ones, that is logical one at the output.

Let's look at the truth table for the OR circuit. Here's the logical OR function. We're going to list the inputs, and see what the output is. We saw 00. No output. 01. Well, if one or the other is open, yes, 10. If one or the other is closed, yes. Output one, both of them? Yes. If both of them are ones, both of them are true, the output is true. That is the logical OR operation, and there's the logic symbol for the OR operation.

Finally, I have to introduce one other simple operation, which is even simpler because it has only one input. This is logical *NOT*, also called *inversion*, because what it does—here's the switch. The switch is now open, and the light bulb is lit. What this particular circuit has is a resistor coming from the battery to the switch, and the switch is placed so that if I close it, it shorts out the light bulb. It provides a quicker path for the current, an easier, lower resistance path for the current, and the current all goes there, and not through the light bulb, so that when I close the switch, and make the input a one, the output becomes a zero, and vice versa. This is the logical NOT operation, then.

There's the truth table. It has only one input, zero or one, and it has one output, which can be either one, if the input is a zero, or zero if the input is a one. That's the NOT operation.

I will remove the batteries, so that we don't keep that light bulb burning and wear out the battery. Now, I want to point out that this particular implementation of NOT is not terribly efficient, because that resistor is dissipating a lot of power, where it's getting warm, and we would not really do it that way, but that shows you how we might go about making a NOT operation.

In practice, it turns out to be more convenient not to make AND and OR, but something called *NAND* and *NOR*. The logic symbol for NOT is this triangle with a little circle after it. If I add that little circle to AND or OR, that becomes logic symbol for NAND and NOR. What they stand for is NOT AND and NOT OR. Logic NAND means a circuit in which the output is one if both inputs are not ones. It is exactly the opposite of AND. It is AND with a NOT added to it.

Here's the truth table for a NAND circuit, sometimes called a *NAND gate*, and this is just a logic symbol. It looks just like the AND symbol. It's got that little circle at the end. The circle means that it is being negated. It is being inverted; it is being NOT-ed. Similarly, here is logical NOR, NOT OR. It looks similarly. It is the symbol for OR with NOT connected to it.

By the way, if I took either of these logic circuits—I'll do it here for NOR—and connected their two inputs together, with A and B, so if A and B had the same value, that would give me either the top or the bottom most row in that truth table, because I had connected them together, so they can't have different values. You'll see what happens. Looking at that, if I have zero at each one of those inputs, I get a one output. If I have one each of those inputs, I have zero at the output, so connecting those two inputs together makes this a NOT gate, or an inverter. You can convince yourself that the same thing would happen for logical NAND; connect the two inputs together, and it becomes an inversion.

We don't really need that NOT operation. We can build it from NOR, or NAND. In fact—I'm not going to go into how to do this—we can build NOR from NAND, and NAND from NOR. We really only need one of these logic circuits implemented in order to make all the logic that we could possibly build in a computer, happen.

How do we actually implement these things electronically? I showed you actual circuits for how we implement them with light bulbs. We can replace switches with transistors. We can replace the light bulb being "on" or "off" with the voltage simply being high or low, battery voltage or zero. That's how we're going to make these circuits with transistors.

Before we do that, let me remind you that I introduced last time this MOSFET transistor, which could be either on or off, depending on the voltage on the gate. There is its circuit symbol. It looks very much like the transistor. The wires at the bottom, the thin lines are the wires connecting the two little N-type regions, and the wire at the top connects to the gate.

How we make these logical circuits with his transistor? Well, there's a circuit. I'll ask you to think for a minute. Is this NAND or NOR. The question is, if input A is high, turns on the transistor, and input B is high, turns on the transistor, what happens? What does the output do?

If you think about it, those two transistors are switches, and if I leave either of them open, there's an open circuit. No current flows. There's no voltage across that resistor, as Ohm's law tells us, and the output voltage is the battery voltage, we've got a one state. If either of those transistors is off, then, the output is a one. The only time the output becomes a zero is when we turn on both transistors, and that is the opposite of AND, so this thing is a NAND gate. There's its truth table that we know.

The symbol for that complicated circuit with two transistors and a resistor is again that symbol, the NAND symbol. By the way, we don't again build the circuit with resistors that waste power. We actually build it with complementary transistors that have the opposite N and P to them, and that makes us able to build a circuit that draws very little power. That's one of the advantages of these MOSFET transistors.

We can also implement NOR with transistors, and you can convince yourself in the same way that this circuit—if you turn on either transistor, the output goes to zero. Only when both transistors are off is the output one, and that is a NOR gate, or a NOR circuit.

I now want to build up in the remaining minutes a more complicated circuit to show you how we get there. There's the symbol for the NOR gate of that, so we aren't going to talk about individual transistors anymore. We're just going to reduce them to simple symbols.

Here is a useful circuit, which I would like to take a minute with and analyze. This circuit has four NAND gates. Let's try to build its truth table over there. Let me remind you—in the upper left—what the truth table of the NAND gate looks like, because that will help to do the analysis.

Let's consider all possible although symbol combinations of A and B at the input here. The first possible combination might be 00. That's the first row of the truth table, so make them both zero. Well, if you look at the NAND truth table at the upper left, if either of the inputs is a zero; the output is guaranteed to be a one. That means that we have a one at the input of two of those two middle gates. If you have a one at either input, and a zero at the other—which we do—the output is a one, so the inputs to the last gate are ones, and that means the output of the last gate is zero. We've got a zero in the output of the whole thing.

Let's go to the next row. Let's put a zero in the upper input A and one on input B. Let's make the same kind of analysis. I'm going to go fairly quickly through this. You can convince yourself this works with the truth table. The output of that gate still has to be a one. Now we have a one and a one on the lower gate. Its output has to be a zero. We have a zero in the one at the upper gate, and its output has to be a one. We have a one and a zero going into the final one, and that gives us a one at the output.

The circuit is clearly symmetric, so we can immediately make the same conclusion for the next gate; and finally, the last case, where we've got two ones at the input means that there's a zero there. We have zeros and ones, which produce ones at the output, and again, we get a zero.

What does this circuit do? What is the nature of this circuit? Well, this circuit could be described in several ways. What is it? Well, one thing it is, is a circuit called *exclusive OR*. If either A or B is true, but not both, you get a true output. If either A or B is a one, but they aren't both ones, you get a one at the output. It's clear from the truth table.

Another thing is that you can think of it as a one-bit binary addition. If you add zero plus zero, you get zero. If you add zero plus one, you get one. If you add one plus zero, you get one. If you add one plus one, you get zero, carry the one. If you built some additional circuitry to take care of that carry, you would be all on your way to building an adder.

Another thing I like to think of it as, is that it's an inequality tester. Its output is one only if the inputs are equal. The question is, are the inputs equal, or unequal? If there are unequal, the output is a one, so it's an equality tester. That's a very useful circuit. It is so useful that we are going to continue this process of building up a hierarchy of more complicated things. We're going to call that *exclusive OR*, or *XOR*, and we are going to give it its own symbol. It looks like symbol for an OR gate, but with this little extra bar at the front that says that this is exclusive OR, and its output is true only if either of the inputs is true, but not both of them. That's exclusive OR.

Now, we can do better than that. We can combine some exclusive OR gates. There are two exclusive OR gates. There's input A1 and B1 to the top gate, and there's A2 and B2 to the bottom gate. Remember that their outputs are true ones only if their inputs are unequal.

There's the truth table for XOR. At the output of this thing, I have a NOR gate as well. Let's look at what happens to this equality tester. The output is one of a NOR gate only if both inputs are zero. What's the only 1 0 both inputs are zero? It's if both outputs of the first two gates, these exclusive OR gates, are zero. When are their outputs zero? Well, this is an inequality tester, so their outputs are both zero only if A1 is equal to B1 and A2 is equal to B2. This becomes an equality tester. It tests for the equality of two binary digits, and sees if they are equal. There at the output, where only if A1 equals B1, and A2 equals B2—A1 is the first digit of binary number A, and A2 is the second digit of binary number A—so that this is a two-bit equality tester. It asks if these are two-bit binary numbers, which could be 00, 01, 10, 11, anything from zero to three, if they are equal.

If you think, "Well, that isn't much," we can very easily expand that. Here's an 8-bit equality tester for the equality of two bytes. It works the same way. We have a great big NOR gate with eight inputs. How do we make that? Well, the same way we made the other individual transistors, although it all mounted on a single integrated circuit. In fact, we mount many, many, many of these gate, maybe millions or hundreds of millions of them on a single circuit—a board nowadays, or a single integrated circuit.

We have a lot of transistors all in parallel like we did for the single NOR gate. There it is, and we have all of these exclusive ORs at the beginning, and we feed A and B, A1 and B1 into the top, A8 and B8 into the bottom. The only way the output of that final gate, that great big NOR gate, can be a one is if all of the inputs are zero, because those individual exclusive ORs are equivalently thought of as inequality testers. The only way we will get a one at the output of this entire circuit is if we have equality at each of those gates, so we get zeros on the output. All of those zeros coming into that NAND gate give us a one on the output.

This is an 8-bit quality tester then, and that begins to be a sophisticated circuit. Eight bits is one byte. One byte is one letter, so when you go through your word processing program, and you want to know if the word "dog" is in there, the first thing that happens is that the letter "D" is a binary symbol, which I showed already is a single 8-bit binary number. That 8-bit binary number against all of the letters it comes to in your document. When you get a one, it goes on and tests it. Then it goes on and tests the next one, and sees if it is an "O." If it is, it goes on and tests the next one. When it finally finds a combination where the first three, D-O-G, are equal to what you have in the document, it has found it. That's how a complicated process like "search function" in a word processing document works; and inside the computer, something like this 8-bit equality tester is going on.

Now, there are many other circuits that execute simple instructions. I mentioned binary adders. There are circuits that shift the whole binary number to the right or left. These serve to multiply or divide by two. Many circuits have more complicated mathematical functions built in, like multiplying great big numbers that include the powers of 10 to them, scientific notation, called *floating point* numbers, but basically, what computers are doing are these very simple, logical operations. Ultimately, these operations we can understand as being built, ultimately, from N- and P-type semiconductors.

Let me end once again with that required picture. Again, we're going from atom to now, all the way up to this equality tester. Here we were at the end of the last lecture. Actually two lectures ago, we have built up to these MOSFET transistors. Then, I showed in the previous lecture how we got much more complicated circuits, but it didn't show you how we hooked them together. Now we're going to get to how we have them together. We've taken that transistor, ascribed a simple symbol to it, built up into gate circuits and here's a NAND gate circuit. We give it a single symbol that now subsumes what that circuit is, and everything that came before it. We built up a more complicated circuit out of that. Here was this tester for inequality, basically, or exclusive OR. We give that its own symbol, so each time, we are building up a more complicated structure of simpler things, and finally, out of that, we built the 8-bit equality tester.

We now have a logical progression all the way from the silicon atom to the 8-bit equality tester that performs a really sophisticated function inside your computer.

Lecture Twenty-Three
How's Your Memory?

Scope: It isn't enough for a computer to be able to test for equality, to add, and to perform other simple operations on sequences of binary digits. It also needs to store the results and input data and to retrieve those stored items for further processing. That's the job of the computer's memory. Memory encompasses relatively slow but long-term storage media, such as magnetic tapes, optical discs (CD and DVD), and magnetic hard discs. But internal to the computer is the fastest memory, based in semiconductor electronics. The ongoing revolution in electronic miniaturization, as embodied in Moore's law, is behind today's computer memories that can store billions of bits of information. In this lecture, we see how individual memory cells work and how they're assembled into today's voluminous computer memories.

Outline

I. Computers use a hierarchy of information storage systems; collectively, they can all be considered forms of *memory* because they "remember" the stored information.

    A. As shown in Module One, optical discs (CD and DVD) provide inexpensive ways to store large amounts of digital information permanently. But they're relatively slow at transferring data, and most can be "written" only once. That's why most discs are considered *read-only memory* (*ROM*).

    B. Magnetic discs, as described in Module Three, store information in patterns of magnetization on a rapidly spinning disc. They have high storage capacity and moderate speed and can be rewritten indefinitely. They retain their information when the power is off. Typically, programs and data are stored on magnetic discs (hard discs) inside the computer until the programs and data are actually in use. Example: A word-processing program and a document normally "live" on your hard disc. When you "launch" a program, it's copied from the disc to faster, temporary memory. When you open a document, it, too, loads into that temporary memory.

C. *Semiconductor memory* is that temporary memory. It's fast, compact, and has no moving parts. It's relatively expensive, but Moore's law ensures that capacity will continue to increase even as cost drops. At some point, semiconductor memory will supersede magnetic disc storage.

    1. Semiconductor memory is *random-access memory* (*RAM*), meaning that any stored item can be accessed as quickly as any other.

    2. Semiconductor memory can be *volatile*, meaning that it loses the stored information when power is turned off, or *nonvolatile*, meaning that it stores information indefinitely. The main memory in a computer is generally fast, high-capacity, volatile RAM; that's why you can lose a lot of work if you haven't saved it (to the nonvolatile hard disc) when your computer crashes. Such devices as digital cameras, removable computer "pen drives," cards for video games, and similar applications use nonvolatile memory. Computers also contain nonvolatile memory that stores the information needed to get the computer started.

II. Semiconductor memory ultimately uses transistors to store and/or transfer stored information.

    A. *Static memory* (*SRAM*, for *static random-access memory*) stores information in the states of individual transistors or logic gates.

        1. Connecting two NOT gates so the output of each gate goes to the input of the other gives a circuit that has two distinct stable states. If one gate's output is 1, then its input, which is the output of the other gate, must be 0. It doesn't matter which output is 1, but whichever is, then the other must be 0. Having both outputs simultaneously 0 or 1 is not possible. Because the circuit can be made to flip from one state to the other, it's called a *flip-flop*.

        2. But how to flip the state? By making the flip-flop from NOR (or NAND) gates, we can control its state. Now we have two extra inputs. Normally, we'll keep them both in the 0 state. As the NOR truth table shows, that means each NOR gate's output will be 1 or 0, depending on whether its other input is, respectively, 0 or 1. We could have 1 at the output of the top gate or 0; either is fully consistent. Whatever value is stored in the circuit stays there.

3. Let's bring the lower input temporarily to 1. If the stored bit is 0, then the other input to the lower NOR gate is 0, and its output becomes 1. Now the upper NOR gate has both inputs 0, so its output—the stored bit—becomes 1. You can convince yourself that if the stored bit had been 1, it wouldn't have changed. Either way, we've ensured that the stored bit is 1. Drop the lower input back to 0, and there's no change, so bringing the lower NOR gate's free input momentarily to 1 makes the stored bit a 1. You can convince yourself that raising the upper NOR gate's free input forces the stored bit to 0.

4. What we have is a 1-bit memory! One of our inputs stores a 1 in the memory and the other stores a 0. We can determine the stored value by seeing whether the output marked "stored bit" is in the 1 or 0 state.

B. The fastest memory is built from flip-flop circuits similar to, but a bit more complicated than, the one we just analyzed. However, this type of memory requires quite a bit of circuitry for each stored bit. For this reason, most computers use a slower but more economical type of memory, called *dynamic memory* (*DRAM*, for *dynamic random-access memory*). In DRAM, each bit is stored as the presence (1) or absence (0) of electric charge on a pair of closely spaced electrical conductors built onto the integrated circuit. (Such an arrangement is called a *capacitor*.) A transistor controls the flow of charge to and from the capacitor, allowing the amount of charge to be changed (writing information) or sensed without change (reading the stored information). Unfortunately, the stored charge gradually leaks away; as a result, this type of memory has to be "refreshed," typically several thousand times per second (hence the term *dynamic*). Although that sounds fast, it's very slow on the time scales at which the computer operates.

C. Whatever kind of memory is used, individual 1-bit memory cells are arranged into 8-bit bytes and then into "words" of typically 4 or 8 bytes (32 or 64 bits). The cells are arranged in an array, connected to logic circuitry that allows individual words to be read (their stored values obtained for use elsewhere in the computer) or written (new information stored). One of the most important specifications for a computer is its total random-access memory (RAM), because that determines the size of the programs it can easily run or the files, such as large digital images, that it can process. A computer with 512 MB (megabyte) of RAM, for example, has approximately 512 million bytes of memory; one with 1 GB (gigabyte) has just over a billion bytes. (These numbers are approximate because computers work with binary numbers, and the size of memory is always an exact power of 2. Thus, 1 MB is exactly 2^{20} bytes, and 1 GB is 2^{30} bytes.)

III. Back to the big picture: We've now gone from atoms to something you can buy and put your hands on—namely, memory for your computer.

Suggested Reading:

Louis A. Bloomfield, *How Things Work: The Physics of Everyday Life*, pp. 478–482.

Marshall Brain, *Marshall Brain's How Stuff Works*, pp. 117–125.

Going Deeper:

Paul Horowitz and Winfield Hill, *The Art of Electronics*, chapter 8, pp. 500–516.

Neil Storey, *Electronics: A Systems Approach*, chapters 10–11.

Questions to Consider:

1. In this lecture, we analyzed a 1-bit memory made from NOR gates. You can also make a memory if you replace the NOR gates with NAND gates. Analyze this memory, and show that you need to keep the inputs normally in the 1 state and set one or the other to 0 in order to store information in the memory.

2. Given that the number of bits in computer memory is always a power of 2, what is the exact size of a kilobyte (kB)? (*Kilo* means 1000.)

Lecture Twenty-Three—Transcript
How's Your Memory?

Lecture Twenty-Three, "How's Your Memory?" I'm not really asking about your memory. I'm asking about your computer's memory. In the previous lecture, we saw how computers can do useful things. They can decide whether two binary bits are the same, or whether two binary bits are not the same, or whether this one is one, and this one is one, or this one is one, or this one is one. They can switch to zero to a one, and vice versa.

Out of that, we build up really useful functions, like comparing two 8-bit words, two 8-bit symbols for a letter to see if they're the same. We can see how that would provide a useful function, say, in a word processing program.

It isn't enough that a computer can simply do those operations, though. It has to be able to put the results somewhere, to be useful. If you're searching for the word "dog" in your computer, and I described last time how that might in fact happen, and you want to change to the word "cat," that's fine, too. You just load the C where there was a D, and so on. Where do you store that, though? You've got to put that information somewhere. Computers need a means of storing their information, and that means of storing is called, generically, *memory*, although we tend to use the word "memory" more for the memory that is internal to the computer itself. We don't think of things like CDs as memory, but they are.

I'd like to talk about all different forms of memory that a computer has, and focus on the conductor memory at the heart of the computer, giving you a sense of how it works. There's a whole hierarchy of ways of storing information in computers. As the technology changes, and as economics vary, the current ones being used vary with time. For example, in the old days, computer memory at the heart of the computer, the so-called core memory, consisted of little tiny rings of magnetic material, with wires going through them in a two-dimensional plane, a whole array of these things. They were painstakingly woven together by hand, to make perhaps thousands of bits worth of memory in an individual computer.

That was superseded, probably, in the late 1970s by semiconductor memory, and we will talk a lot about that, but computers also have other forms of memory that tend to be slower, harder to access, but able to store more information, and be cheaper. It's that equation: the economics of memory, the amount of material it can store, and the speed, are all trade-offs; and that is continually changing. What's true of computers today then, and what kind of memory they have, will not be true tomorrow.

Ten years ago, computers didn't always have CD drives, because optical storage was a new thing. Now, all computers have CD drives, and probably most have DVD drives. They can store information on DVDs, which as we know will hold a lot more information than CDs will, and we understand why. It has to do with the diffraction limit, and the wavelength of the light used, and so on.

I'm going to give you just a quick overview of the kinds of memory that a typical computer might have at the time we're making this course, and that is going to change as time goes on. Typically, there's some kind of external storage. Let's look at a picture of a typical desktop computer here, with a tower-type configuration. At the top, it has a slot for an optical drive—CD or DVD. Eventually that will also be Blu-Ray as well, I'm sure. There's a recordable, removable medium of some sort, a CD in this case, a writable CD on which I can store information, or on which I can have stored information brought from elsewhere, and enter that into my computer.

Now, a CD can store about 640MB of information. A DVD can store something on the order of not quite 5GB, 5 billion bytes of information. That's a lot of storage, but that storage is relatively slow to read. You put a DVD or CD in your computer, especially if it's got a movie on it, it can actually take a long time to read out, so long that the movie kind of appears on your screen haltingly.

Optical storage then, on these plastic disks, CDs and DVDs, is cheap. CDs cost way under one dollar. You can get them for $0.10 apiece if you buy them in bulk, cheap. They're easy to write to, easy to read, but slow, relatively slow. It's a high-volume storage medium, but slow. The advantage is that it's external and it can be removed. You can archive information and store it somewhere else.

The workhorse of your computer in terms of long-term storage of information is its *hard disk*. Almost all computers have hard disks. Hard disks are ultimately magnetic. The information is written as little domains of magnetic material, and similar to the way I described earlier, it writes a VHS tape or an audiotape by passing a current through a tiny coil that's near the magnetic medium. The magnetic medium moves past the coil, and imposes a pattern of magnetization on it. It used to be that that pattern would be read, as I described a few lectures ago, by electromagnetic induction, as the discs spun. The different, changing magnetic regions would induce a current, and that would take the information off the disc, the so-called *magneto resistive effect* is used. What that does is to change the electrical resistance of a little sampling element in the head as the disc spins underneath. That has the same effect to produce an electric current. It varies in a way that corresponds to the information stored in the disc.

The second major piece of memory that a typical computer has is its internal hard disk. It also might have an external hard disk, which you can connect for extra storage, for backup, whatever. I have an actual hard disk here. It's out of a computer that's a few years old. This hard disk is the size of the disc that would be inside a desktop computer; a laptop computer would have a disc that would be maybe one fifth of this size, a tiny little disc.

Inside here is the disc itself. I will take it apart in a minute. On the bottom is the bottom end of a circuit board that has a lot of integrated circuit chips on it whose sole job is to read and write information from the disc, and translate for that string of bits that's coming off the disc into a form that can communicate with the rest of the computer.

In the case of computer hard disks, there's a platter of metal, coated with a magnetic material, and it's spinning at very high speeds, thousands of revolutions per minute, and there is a head that literally flies, held aloft, if you will, off the disc by aerodynamic forces. The distance is on the order of one millionth of one meter. A disc crash is literally a crash. It's a little bit like an airplane crash; that flying disc—something happens, it hits a particle of dust—loses those aerodynamic forces, and will crash into the disc, and damage the disc surface, damaging the head, and there goes your Ph.D. thesis if you haven't got it backed up, or all of your pictures of Grandma, whatever you had at that point on the disc. If it wipes out the directory of the disc, which is what tells the disc where everything is, even though everything else is still fine, you may have lost it. That's why this hard disk has a cover on it. I'm going to take the cover off, and show you the hard disk.

This hard disk is actually a stack. You can't see it very well, but it's actually a stack of four or five of these shiny metal disks. They spin very rapidly, and there is a small projection here, which is the head, and it can move in and out, back and forth, in an arc. It can't do that right now, because it's locked, but it can move in an arc, and read and write information off of the tracks on this disk.

This is an electromechanical device. This disk is spinning at very high speeds, and this is where the information is stored, so that when you turn on your computer, part of the noise you hear is the hard disk spinning, because in a desktop computer, it tends to spin all the time. In a laptop computer, especially if you are on battery power, it may spin up and down to save on battery, but that obviously makes it slower to get information.

It's easy for this head to move back and forth. In fact, you will hear clicking noise as your computer boots up, or as you read a big file off the disc. That's this head, physically, mechanically, moving back and forth to different parts of the disc, to read the information that is stored in different places. As the disc gets more and more full, a single file—your term paper, your Ph.D. thesis, your pictures of Grandma—may be stored in little hunks in different parts of the disc, and the head has to go around and find all of those, and put them together to make a complete file.

What we put on hard disks? Well, we tend to store our documents, things that we've written, pictures that we've taken, songs, music, whatever, all kinds of things—programs, those kinds of things, tend to be stored on hard disk. In addition, we store software on hard disk. When you double-click your word-processing program, your word-processing program lives on the hard disk most of the time. When you double-click it, it's loaded into the computer's internal memory, which I'll get to in a moment, and it is executed from there, because that's much faster. Its permanent location though, is on the hard disk. When you hit the "save" button on your computer, or if you use the save command, what you are doing is taking information that's stored in the computer's temporary but very fast internal memory, and writing it onto the hard disk.

The beauty of the hard disk is that it doesn't have to be spinning to remember. When you shut down the computer and turn it off—the hard disk—it's there with its regions of magnetization, and they don't go away any more than when you take your favorite movie out of the VCR, and it ceases to be on the videotape. This is just like videotape or audiotape. The information is stored there magnetically, and it is still there. It's called *nonvolatile memory*. It doesn't disappear when you turn the power off.

The hard disk, then, is nonvolatile memory. It's easy to get very large hard disks. At the date on which I'm recording these lectures—and again, it's hard for me to talk about timing with computers, because of Moore's law, and everything doubling in 18 months, but computer hard drives are approaching 100GB, 100 billion bytes of information on a good-sized computer hard drive, and you can certainly by hard drives that are much bigger. A thousand GB, a terabyte, used to be the province of supercocomputers only. Now, you can buy a terabyte of storage. If you're something like a photographer with zillions of digital photographs, you might need a terabyte of storage. Some of us can get by with 20, 30, 40GB. With computers of 10 years ago, it was 20, 30, or 40MB, a factor of 1,000 increases since then. There's Moore's law at work.

The hard disk storage is relatively cheap, although the hard disk is a fairly complicated mechanism. It's an electromechanical mechanism. It's a very delicate, sensitive mechanism. It's cheap only because you can get huge, huge, huge amounts of storage crammed into a relatively small space at relatively small cost, but it's not terribly fast, because again, you have to move that head physically and mechanically, a process that takes a fraction of a second to get it to the right place, and then, as the rotating disc goes around, you have to wait until the right information is under the head to get it off. The hard disk is therefore not fast.

They're pretty reliable these days, but not perfectly reliable, and you ought to back them up, because that hard disk, with 100GB, or whatever, is holding an awful lot of valuable information. If you lose your hard disk, and you've got no backup, you've lost an awful lot.

Then, there's a third type of memory inside the computer, and that's the computer's internal *random access memory*, or *RAM*. I've got some RAM here. I showed you this in the last lecture, too. This is a RAM module. In this case, it's got nine individual memory modules on it, each with a few hundred megabits of information. There are nine of them. Eight of them are the eight bits that make up a byte, and the ninth is for processing and error checking purposes. I will talk more in this lecture about how that random access memory works, but this is random access memory.

A hard disk is somewhat random access. A tape is called *sequential memory*, because you can't get information off a tape until you have wound the tape all the way to the point where the information is. A hard disk is pretty random. Once you get to the part of the disc you want to be on, you have to wait until the right bits come around. It's kind of the same with a CD, or DVD, but this memory is truly random access. You can get to any point in the memory. You can get any information out in an equal amount of time. You don't have to wait for it to go past other amounts of memory.

We will talk about how that memory works in just a minute, but today's computers typically have a substantial amount of computer memory, of

RAM memory, typically—well, my laptop here has a gigabyte. That is still probably on the large side. This is because I use this for scientific research, as well as normal computer uses that everybody uses. Five hundred MB of memory would be a typical amount of RAM, with 256 getting a little low. You may notice that those numbers tend to be kind of strange numbers, and the reason is that those numbers are powers of two. I will show you in a minute why that has to be.

Semiconductor memory, memory made of semiconductors, ultimately of the kinds of transistors we've been talking about, is the fastest. It's compact, it has no moving parts, and the only reason it isn't used as the dominant memory computers is because it's more expensive. Well, partly that, and partly that the best and fastest semiconductor memory is also volatile, meaning it stores information only while the power is supplied, and if you take the power off, it's gone. That's why, if you forget to save your document, and your computer crashes or the power goes down, you've lost, because everything that was in the computer memory is totally wiped out and gone. You've got to hit that save button frequently, or you've got to program the auto saves, and if there's a crash, you can bring you back everything that was there, at least until the last time it saved information.

It is possible to get semiconductor memory that's nonvolatile. I have an example of it here. This is my little pen drive. It's actually called a *drive*; because it used to be that I would buy a small hard drive, to store excess information on with my computer. In recent years, semiconductor memory has become cheap enough that we can build these little tiny pen drives. This is a 256MB drive. It has space for one quarter of a gigabyte in information that can be stored on it. I simply plug it into the back of my computer, and it becomes available as another storage medium.

It consists of semiconductor memory that is nonvolatile. The trade-off for the nonvolatile memory is that although it retains its memory when the thing is off, it's a lot slower, so that getting information on and off of it is not nearly as fast as getting it on and off my internal memory. By the way, the memory in a digital camera is the same type that's in that little USB pen drives that I have there.

Let's talk about semiconductor memory now, and how it works, because we're trying to get from an empty computer, and we want to relate this to what we know about atoms, semiconductors, PN junctions, transistors, and so on.

Well, we've built up beyond that. In the last lecture, we introduced the symbols for logic circuits, and said that we weren't going to think about individual transistors. They're in there. We understand that they're in there, and how they work. We can go all the way back to atoms, but we're going to start with these symbols.

Here's a very simple memory built from a couple of NOT gate. This particular memory is called a *flip-flop*, and the reason is called a flip-flop is because it only has two possible states that it can be in, only possible configurations. Let me show you.

Suppose the output of that right hand NOT gate is a one. Remember what a NOT gate does. A NOT gate's truth table is if the input is a zero, the output is a one. If the input is a one, the output is a zero. It is an inverter, a NOT-er. It takes the input, and makes it whatever it wasn't.

Well, these two NOT gates are connected together in a funny way. The output of the right hand one goes around through that wire from above, and goes in to the input of the left hand gate. The output of the left-hand gate goes into the input of the right hand gate. These two gates, these two logic NOTs are connected together, the output of one to the input of the next, the output of that one back into the input of the other one.

What's going on? Suppose this is a one. That means it in a typical computer circuit, it that a certain voltage. Five volts is often the value. Zero volts would correspond to a zero, and this is actually quite a range of voltage, all of which will be interpreted as a one, and all of which will be interpreted as a zero. That's what makes logic circuits digital, and on-off circuits so robust. It's hard for them to have mistakes.

There's a one at the output of the right hand gate, and that means that that one, because the wire that connects is at the input of the left hand gate, and that the input of the left hand gate is a one. It's a NOT gate. Its output is zero. If its output is a zero, that's the input to the right hand gate. The input to the right hand gate should be a one. This is a perfectly logical situation for this circuit to be in.

It's a one by memory. Right now, if we think of the right hand gate's output as the state of the circuit, this circuit is in the one state. It's storing a digital one. It is storing a "yes," a one, a binary number one.

It's also possible for this circuit to be in a different state. The output of the right hand gate could be zero. Then, the input of the left hand gate is also zero, because they are connected together. Well, an inverter, a NOT circuit flips that to a one. It puts a one at the output of the first gate, the left hand gate, and a one at the input to the right hand gate should make a zero at the output. That situation is also perfectly logical and allowed.

There is a situation, though, that is impossible. It is impossible for the output of both gates to be one. Why? Because that's inconsistent with what these NOT gates do. That's logically inconsistent. That's not possible.

What I have in this 1-bit memory is a circuit that can exist in one of two possible states, and that's all, and once it's in of those states, it will happily store that information, as long as you keep the power supplied forever. The trouble is, I can't easily change the state of that circuit. So to show you how we can change the state, I want to draw that circuit in a slightly different way.

Here is the same 1-bit memory circuit, drawn in a different way, a more symmetric way. The two gates kind of enter symmetrically. Just convince yourself for a minute that that circuit is exactly the same as what I had before when I started this thing out, two NOT gates connected, one output to the input of the next one, and so on. That is exactly what I have again when I do this, when I draw it like that. It's exactly the same circuit, just drawn differently and emphasizing the connection where there are wires crossed at the beginning. The output of the upper gate going to the input of the lower gate, and the output of the lower gate going to the input of the upper gate.

Now, I want to make a circuit in which I can change the state of the circuit somehow easily. I can say, "Okay, let me put a zero in there, because I determined that the result of that operation was a logical zero, and that's what I want to store in that place." How do I change the state of this simple memory circuit?

Well, I will replace those NOT gates with a more complicated gate that has an extra input. In this particular case, I'm going to replace it with more gates. Again, convince yourself that that circuit and that circuit are essentially the same, except that I have those extra gates in place, and I'm going to call the output of the upper gate the *stored bit*, and it could have the value of one, or the value of zero. That's what I'm going to do there.

Now, I'm going to figure out how I can change the state of the circuit. Before I do, let me point out that what we typically do with this circuit is to normally keep both of those extra inputs in the zero state. If we do, if you look at the truth table, or the NOR circuit, at the lower left, if one of the inputs of a NOR gate is a zero—that would be the first row, the second row, the third row—the output could be either a one or a zero, depending on what the other input is doing. In the first row, if one of the inputs is a zero -- if they're both zero. If the output is a one, the first input is a zero, and the second one is a one, the output is a zero, and so on.

The point is this. If we are in the zero state at both of those inputs, this circuit basically behaves like the other circuits I just had. These gates simply act as inverters. If the first one is a zero, if one of the inputs is a zero, the output is the opposite of what the other one is. Take a look again at the first two rows in the truth table. If the first input is a zero, input A is a zero, then, whatever input B is, the output is the opposite, so those two gates are behaving just like those converters are NOT gates that I had before.

If we have those two extra inputs in the zero state, then this circuit is exactly like the circuit we had before. Nothing new and exciting happens. Let's do a little analysis. Let me remind you that there's no connection again, so here we are with these two extra inputs in the zero state, and let's suppose for a moment that the stored bit is a one. Let's convince ourselves that this is all consistent. If the stored bit is a one, because the output is connected to the input of the lower gate, that input is also a one.

Take a look at the truth table. It's a one at the upper input, A, a zero at B. That's the third row in the truth table, and that says the output should be a zero. If the output of the lower gate is a zero, that puts the other input to the upper gate at zero. Both inputs to the upper gate are zero. Well, that's the first row in the truth table. 001. The output of the upper gate should be one. It's all totally consistent, and everybody's happy.

You can convince yourself in the same way that if the stored bit were a zero, there are two zeros on the lower gate. We are in the upper row, and the output of the lower gate is a one. It puts the input of the other upper gate at one, and a zero and a one inputs gives us a zero at the outputs. That's perfectly consistent as well. This device can be in either of these two states, and everybody's perfectly happy. It will simply stay in that state as long as it needs to, which is basically forever, as long as the power is applied.

However, now let's try to change the state. Let's take one of those new spare inputs, and temporarily change it. Let's change the lower one, for example, to a one. There it goes, turning to a one. Now, we look at the truth table, and the truth table says that if we have a zero and a one, we ought to have a zero on the output, so that output of that lower gate has to change the zero. Now, we have a zero, and a zero on the upper gate, so we look at the truth table again, and we see that a zero and a zero gives us a one at the output. That means that the stored bit has to become a one.

Then, we have a one and a one at the lower gate, and that requires that our output be a zero, which it was, so that was fine. Now, if we bring the lower gate back down to zero, that still gives us a zero on the output of the lower gate, and so we're still perfectly consistent. What we did by raising that lower input to a one temporarily was to change the stored bit to one. Then, when he brought that lower input back down to a zero, the stored bit stayed a one, and we stored a one in there.

That is the way we change the state. I'm not going to go through all of the analyses, but you can convince yourself that if we do the same thing with the upper loose input, and changed it temporarily to one, we would force output to be zero, regardless of whether it had been one or zero before. If it had been zero, it would stay zero, and if it had been one, it would go to zero. Then, we would have stored a zero as the stored bit, and then, when we dropped that input back down to zero, the zero would stay as the stored bit.

Now, we have a circuit in which we can change what's happening. I would like to think of this circuit more as a little memory circuit that can store a zero temporarily if I raise the top upper input to a one, or temporarily store a one if I raise the lower input to one. Don't you dare raise them both to one simultaneously—that gives inconsistencies.

Real memory circuits are more complicated than this, to avoid that kind of inconsistency, and some other timing inconsistencies that can occur in these simple circuits. Real memory circuit is a little bit more complicated.

Well, some are really a little bit more simple. This kind of memory I've just described, built from flip-flops, is, in fact, the fastest kind of semiconductor memory we can make. Some computers, like supercomputers, use semiconductor memory of this sort—flip-flop based memory. Most of our personal computers use what's called *dynamic* random access memory. The kind I have just showed you, with the flip-flops, is *static* random access memory. Once you've stored the bit, it stays there.

In dynamic random access memory, there's only a single transistor for each bit. It dumps charge onto, or reads charge off of, a pair of closely spaced electrical conductors called a *capacitor*. The problem is that the capacitor loses its charge on a timescale of less than 1000th of a second, typically, and so, a computer with dynamic random access memory, remember, you have to go through a few thousand times a second and refresh this memory. That may sound very fast, but computers operate one billion times per second in their basic operations, so having to refresh the memory a few thousand times a second is a very, very long time interval, as far as the computer is concerned, so that's not really a problem. Your computer memory is probably dynamic random access memory, and it works slightly differently from this flip-flop based memory that I've showed you here. This kind of memory though, is, in fact, used in computers as well.

Whatever kind of memory is used, this simple circuit, now, is going to reduce—whether it's this circuit, or a dynamic random access circuit—it to a single little cell. We add a few more gates to it that allow it to do the following. We've got an output that corresponded to what are stored bit was—whether it was a one or zero—and that's the stored bit. We have an input where we put in what bit we would like the output to become. We have a *read/write* control that tells us whether we're going to write a new piece of information to this cell, and change its output from a zero to a one, and changed its stored bit from a zero to a one, or whether we're going to read out the stored bit. We can either write to memory cell then, or we can read out from the memory cell. The select line at the top says, "Okay, if I make that high, if I make that a one, this cell is active. It's the one that's going to get read from or written to. If the select line is low, even if there are things connected to the other inputs, nothing happens, because this cell isn't selected."

That's a single, individual piece of computer memory. Well, when we go from there? We take those individual cells and we build them up into a rectangular array of memory cells. Each of them has its read or write input all interconnected together, so that when we put a logical one, a high-voltage five volts on that read/write line, all the cells are ready to have information saved to them, if "one" means write, and a zero means read, but only the ones selected are the ones for which that will happen, and you'll notice the select lines, those are the lines at the top, that are connected together for the top four memory cells, they're connected together for the next four memory cells, and they're connected together for the third row down, and the fourth row down. They all go to this block on the left marked "address decoding logic." In this particular case, it has two input lines at the bottom, and two input lines, it's a two digit binary number that could be any one of four values, 00, 01, 10, 11—decimal would be 0, 1, 2, or 3, four possible values—and those correspond to the four possible arrays of memory cells along here.

That first one is the first word. These are 4-bit words, as opposed to the 64-bit words used in today's modern computers, with 64 of these things going along. That's word two, that's word three, and the last one is word four. Depending on what comes in on those address lines, those 2-bit address lines, one of those words will be selected as the one to which the read/write operation will occur, and depending on whether the read/write is set to read or write. If it's set to write, whatever is on those four input lines—a 4-bit binary number, half a byte—will go to that particular word.

If we are asking for reading, then whatever is in that particular 4-bit word will appear on the output lines that I showed at the bottom.

That's how an actual computer memory is made; and nowadays, if you buy a piece of computer memory, it doesn't have 1, 2, 3, 4 words, it might have 500 million words, and it has more bits coming in on the binary addressing, but any of those bits can be addressed equally well. That's what you are buying when you're buying computer memory. When you are buying one of these simple little modules and installing it in your computer, you are installing an array that looks like I've shown in that picture, an array of memory cells, with a very, very large number of cells, and the computer sends a binary number to those address lines, and selects which words are getting read or written. And when you double-click your Ph.D. thesis or your picture of Grandma, or whatever it is, what happens is that thing is loaded into computer memory, and then, when the program that you're working with needs to read it, it sets the memory to read, finds the right addresses where that information is being stored, and it reads and processes it as you want it processed, then it writes it back out, perhaps in modified form, with the right lines.

Again, one of the most important specifications for your computer is how much memory it has, because if it runs out of memory, it can't do something like process that big picture of Grandma that you took with your new six-megapixel camera, which has many, many, many, many bits of information. Next time, we will go through how digital photography works a little bit, and we will see exactly how much information is stored there. You need a lot of memory to run big programs, and that is why memory is so important. The amount of RAM you have in your computer is such an important specification, and again, these days, a few hundred megabytes, probably, and a few gigabytes a year or two from when I make this tape, that's the kind of memory you'll be having. Your hard disk is holding perhaps 100 times as much as that. I have a lot more in your hard disk than in your RAM. Sometime, probably in the next 10 years, semiconductor memory will become so cheap because of Moore's law that it will be cheaper to make semiconductor disk drives, which we've already done for these portable units, and we will say goodbye to the electromechanical hard disk.

Let me wrap up, again, with the hierarchical picture. Computer memory is all put into these little memory modules. I showed you these modules, and we're ready to go from atom to computer memory. We already went from atom to simple logic circuit, like an exclusive OR gate. Now, we've put together some of those simple logic circuits to make memories. We've combined a little more logic to make memory cells. We've joined those together to make an array of memory cells, and have packaged them onto one of these little strips, with a bunch of integrated circuits. That's the memory module you buy and stick in your computer for extra memory.

Lecture Twenty-Four
Atom to Computer

Scope: With one example of a processing circuit (the equality tester, Lecture Twenty-Two) and knowing how computer memory works (Lecture Twenty-Three), we're ready to put it all together to make a complete computer. Computers aren't just about numbers, however; they also process information, such as text and images. For this reason, we also need to understand how such information is coded into the binary language computers can understand. And we need to be able to get information into and out of our computer. Here, we look briefly at the overall structure of computers, how they encode text and other information, and their input and display devices.

Outline

I. A complete computer comprises a processing unit; memory, including high-speed semiconductor RAM, a magnetic hard disc, and drives for optical discs; input devices, such as a keyboard and mouse; interfaces for connecting to the Internet; and a video display.

   A. The *central processing unit* (*CPU*; often called a *microprocessor* in personal computers) is at the heart of the computer. This unit contains circuits, like the equality tester we developed in Lecture Twenty-Two, that perform simple arithmetic or logic operations on binary numbers (the 32- or 64-bit "words" used in today's computers). The CPU also has limited but very high-speed temporary memory, called a *cache*, for storing instructions to be executed and data to be operated on. There are also special memory registers that keep track of the steps to be performed in a program, circuits that fetch and decode specific instructions, and a *clock* that generates electrical pulses at a regular rate to keep all the computer's actions happening in an ordered sequence.

      1. When you hear that a computer contains a "G5 chip" or a "Pentium chip," that's a description of the specific model of microprocessor used in the computer.

2. The speed of the computer, usually expressed in gigahertz (GHz, or billions of Hertz), describes the rate of the CPU's internal clock; 1 Hertz is one clock pulse per second. Typically, a computer can perform the simplest tasks, such as fetching a word from memory or adding two numbers, in one to a few clock cycles. Thus, a 2-GHz computer (2 billion clock pulses per second) might take as little as half a billionth of a second to perform a simple task. Equivalently, it might perform as many as 2 billion such tasks each second. However, some CPUs have more complex circuitry that lets them perform several operations at once; for this reason, the clock rate in GHz is not a reliable indicator of a computer's overall speed.

B. The CPU is mounted on a *motherboard* that also contains the random-access memory and other integrated circuits. The board itself is made of insulating material like fiberglass or epoxy, with electrically conducting copper pathways connecting the integrated circuits. Modern motherboards are multilayer structures, allowing complex interconnections among the various integrated circuits. The motherboard is essentially the complete computer, minus long-term memory devices, such as hard drives, and input and display devices.

1. The motherboard is characterized by the speed and width of its *data bus*—the name for the channel that carries digital data back and forth between the CPU, memory, and other devices. A 64-bit-wide bus carries 64 bits simultaneously. A modern high-performance data bus might have a width of 128 bits and can transfer hundreds of millions to a billion or more 128-bit chunks of data each second.

2. Also on the motherboard are specialized integrated circuits that transfer data to and from hard discs, CD and DVD discs, and other data storage devices; circuits that communicate with input devices, such as the keyboard and mouse; circuits that format data and send it to output devices, including displays and printers; general-purpose interfaces, such as the universal serial bus (USB); and circuits for high-speed network connections. The motherboard usually has empty slots for additional memory or specialized purposes, allowing the computer to be customized.

3. Communications between the motherboard and the outside world can be either *parallel* or *serial*. In parallel systems, large numbers of bits are transferred all at once. Parallel communication is very fast, but it requires large numbers of wires. In serial communication, bits are sent one at a time. Most peripheral devices, and all network communication, use serial data transfer.

C. Input and output devices permit communication between the computer and a human being.

1. A computer's keyboard is simply a set of switches arranged in a grid. Pressing a key causes electric current to flow on two wires that identify the key's horizontal and vertical locations in the grid. A built-in microprocessor identifies the key and transmits a code for the associated letter or other character to the computer. The microprocessor also identifies special key combinations, and it eliminates multiple keystrokes that might occur because of mechanical "bouncing" of switch contacts. Some keyboards use so-called capacitive switches, in which there is no actual electrical contact.

2. The first computer mouse was a simple device in which a rolling ball turned two perpendicular shafts whose rotations were detected optically. The mouse used most often today is entirely optical, bouncing light from a light-emitting diode off the surface on which the mouse is moving. The reflected light is detected by a sensor similar to that in a digital camera, and a tiny built-in microprocessor analyses the changing patterns to calculate the mouse's motion. That information is sent to the computer, which uses it to move the cursor on the display device.

3. Modern computers generally use *liquid crystal displays* (*LCDs*). In these devices, application of a voltage to individual picture elements, or *pixels*, alters the orientation of the molecules in a so-called liquid crystal. This, in turn, changes the polarization of light coming through the crystal and either blocks or allows passage of the light through a subsequent polarizing filter. Red, blue, and green color filters at each pixel allow combinations of colored light to give a full-color image. Touch-screen displays used in supermarket checkouts, information kiosks, and the like include a surface

layer that detects the location of a finger touch, allowing the display to function as both an input and output device.

II. Computers don't just compute: They process all sorts of information, including text and images. Ultimately, though, that information is encoded as binary numbers.

    A. Each text character is encoded as a single 8-bit byte, as we saw briefly in the previous lecture. For example, the capital letter A is assigned the binary equivalent of the decimal number 65 (01000001); a blank space is the number 32, or 00100000 in binary; and the number 8 (00001000) is the backspace character. The word *text* is a sequence of four 8-bit bytes: 01010100 01000101 01011000 01010100. A document you just typed, or even the whole book you just read, has a similar representation as a sequence of binary bytes. (This coding is called the American Standard Code for Information Interchange, or ASCII; if you save a computer document as "ASCII text," you're storing just these codes, without a word processor's special formatting, such as bold or italic lettering.) Often the codes are written in base 8 (octal) or base 16 (hexadecimal) because they translate easily to base 2.

    B. The binary representations of music and images require huge amounts of binary information. Each pixel in a digital image, for example, must be encoded as binary numbers representing the intensity and the color of the light. *Compression* techniques reduce the number of bits required without significantly altering the quality of the information. For example, movies can be compressed with the MPEG (for Motion Pictures Experts Group) scheme, which finds redundant pixels both within a given frame of the movie and in adjacent frames and eliminates them in the stored data. The much smaller MPEG files are easily stored or transmitted over the Internet. The compressed images are "decompressed" to restore the missing information for display. Similarly, digital cameras often use JPEG (Joint Photographic Experts Group) compression, which can reduce a 15-million-byte image to fewer than 1 million bytes. And the notorious MP3 music format is really a compression scheme that extracts music from a CD and reduces the number of bytes by a factor of about 10. That's what allows easy swapping and storage of music files, much to the consternation of the recording industry. (MP3 is actually the audio portion of the MPEG movie compression scheme.)

III. Wrap-up of Module Four: We really have come all the way from atom to computer!

Suggested Reading:

Marshall Brain, *Marshall Brain's How Stuff Works*, chapter 5.

Henry Fountain, *The New York Times Circuits: How Electronic Things Work*, pp. 36–77.

Going Deeper:

Paul Horowitz and Winfield Hill, *The Art of Electronics*, chapter 10.

Neil Storey, *Electronics: A Systems Approach*, chapter 12.

Questions to Consider:

1. Look up the specifications for your own personal computer, finding (a) total RAM in megabytes or gigabytes; (b) hard-disc capacity in gigabytes; (c) microprocessor type; (d) clock rate in gigahertz; (e) bus width in bits; and (f) bus rate in megahertz or gigahertz. Explain what each means and what role it plays in establishing your computer's performance.

2. Suppose you have a 5-megapixel digital camera, meaning that the image sensor records information from 5 million different spots. In a typical camera, the intensity of each of the colors red, green, and blue at each pixel is encoded as a 24-bit binary number. Estimate the size, in bytes, of a "raw," uncompressed image from such a camera. Remember that there are 8 bits in a byte. No wonder compression is needed to get manageable file sizes!

Lecture Twenty-Four—Transcript
Atom to Computer

Lecture Twenty-Four, "Atom to Computer," has almost the same title as the whole module, "From Atom to Computer," because here we're going to put it all together. We now know how to build some of the basic processing circuits that go on in computers and we can trace all the way back to a silicon atom a general understanding of how they work.

We don't yet know how we put that all together to make a complete computer that computes and does all the things we want it to do; and so, the purpose of this last lecture in Module Four is to do all of that together.

What does a complete computer consist of? Well, it's got a processing unit, an integrated circuit, or sometimes several integrated circuits together. They're closely connected; usually, in a personal computer, there's just one integrated circuit, called the *central processing unit*. It's also called the *microprocessor*, and the microprocessor increasingly means any of the tiny little specialized computers that are built into things like cell phones, CD players, or whatever else.

There's a central processing unit then, and that's the real part of the computer. It's got circuits like the equality tester that we looked at last time. It's got special circuits for adding binary numbers and circuits for comparing binary numbers. It's got special very high-speed memory built right onto the chip itself, onto that integrated circuit memory called *cache*, whose purpose is to store instructions in the program that the computer is executing. Maybe it stores a few instructions ahead of where you already are so that when an instruction is ready to be executed, it's already right there in the central processor, and the central processor doesn't have to go out to memory to fetch it. So, if you hear of a computer having something like "level two" cache, or level one cache, or however big it is, you're talking about that special memory, very high-speed memory that's built right in onto the processor chip.

It's also got a number of other circuits. It's got special memory registers that keep track of what steps are to be executed in the program. Specifically, circuits that decode instructions, because the instructions themselves that tell the program what to do—the instruction that says, "Add," "Go to memory register one," "Get the word that was in memory register two, and put it back in memory register three"—that instruction is itself ultimately just a binary number, which is not being interpreted as a number. It is not being interpreted as text. It is being interpreted as an instruction to the computer to do what it's supposed to do; and so, when that instruction comes in, it has to go to the circuit that decodes it, and actually makes the computer do with that instruction says it ought to do.

By the way, one of the worst things that can happen in computer programming is to have data, like your Ph.D. thesis, or Grandma's picture, *overwrite* what were supposed to be instructions. Some of the worst computer crashes you will have are caused by that kind of thing, where the computer overwrites instructions, and then the instructions aren't there to be executed, so that the program can't do its job correctly. Good programming avoids that, but the complexities of modern computers and modern programs means that that happens not infrequently.

One of the most important functions on the CPU, central processing unit chip, is the clock. The clock is simply a very simple circuit. It's very closely related, actually, to that flip-flop circuit a shot in the previous lecture, except it's a flip-flop that's never stable in one state, so it flips back and forth between two states at a steady rate. That clock produces a series of electrical impulses, again, at a steady rate, and they're sent throughout the whole computer. Their job is to keep all of the complicated things the computer is doing synchronized, in step with each other, so that the computer doesn't just randomly add some numbers, and then do this, and then do that. It does everything in sequence. The clock pulse comes along and that's the time information gets loaded into memory. The next clock pulse comes along, and that's when an instruction gets executed, and so on and so forth. In modern computers, several things may be going on at once, but there is a basic underlying clock that produces these pulses at a regular rate, and that ultimately, in some sense, determine what's going on, and what the speed of the computer is.

You may hear that your computer has something like a Pentium chip. Here's a motherboard from a computer with a Pentium chip, for example, or a G5 chip in the case of Apple's computers, but it's made by IBM, often. Those chip names are the model of a particular kind of central processing unit, and after the first of such a model is introduced, there may be many variants on that same model: Pentium II, Pentium III, G5, and then a G5 that will come out small enough for a laptop computer, and so on. Names like Pentium and G5, then, refer to the type of microprocessor chip, or central processing unit chip you have.

The speed, which these days is usually measured in gigahertz—a gigahertz is one billion clock cycles per second—is a measure of the speed at which that underlying clock is producing electrical pulses. Click, click, click, click. If you have a 1GHz computer, it's one billion times per second, if you have a 2GHz computer; it's two billion times per second.

Very crudely—and I want to emphasize that this is crude—that clock rate is ultimately what determines the overall speed of the computer. I remember my first computer. It had clock rates of about 8MHz, eight million cycles per second. Now, we're measuring clock rates in gigahertz, but a word of caution. Different models of chips have different numbers of processes going on at once, so a 1GHz chip from one manufacturer may be the equivalent in speed to a 2GHz chip from another manufacturer. All other things being equal though—namely, the basic architecture of the chip, and how it's put together—a faster clock rate means a faster chip; but, when you are comparing among different manufacturers or other chip types, be careful about that. Two GHz isn't always twice as fast as 1GHz.

What that means is that the clock rate ultimately sets the timescale for the very simplest operations a computer can do. For example, in one clock cycle—say, one billionth of a second for a 1GHz clock, or a half a billionth of a second for a 2GHz clock—that's the time it takes to execute the simplest things, like fetching a number of memory, and loading it into the central processing unit ready to process, or adding two numbers, or comparing two numbers. Those simple operations will typically take one to several clock cycles, so that a 2GHz computer can perhaps do two billion operations a second. That's fast. That's a personal computer. Some of the world's supercomputers are far faster than that, although they have a slightly different measure for how fast they are actually going, called FOPS, Floating Operations per Second. We won't go there, though.

Now, the CPU chip is not all by itself. Again, here's a motherboard, from a 1990s vintage computer. There's the big Pentium central processing chip, and it's mounted on a board called the motherboard. The board is typically made of fiberglass. It's an insulating board. I've described this process before. It's been coated with copper, and then the copper has been photo exposed, similar to the way one integrates circuits themselves, but on a larger scale. The copper is then etched away where it isn't protected by photoresist, and that makes the various interconnections. The Pentium chip and all of these other smaller chips that perform various functions are connected among each other by wires on both sides of the board. The back of the board is mostly a single piece of solid copper, but on the front side of the board, there are many different little pieces of wiring, little pieces of copper running among these chips.

That's what a motherboard looks like, and a motherboard has a number of things on it that are useful. Think about. For example these large black structures here are connectors where cables come in that connect other parts of the computer. Over here, in these white bins, are places where you would install memory chips, or where memory chips would have been installed, if they had not been removed from this. If you were to look inside your own computer, you might find a number of empty slots like this. If you buy additional memory, you could probably figure out how to install it, and then you would have more memory in your computer.

Often, the motherboard is built with some empty sockets that are designed for expansion, or modification, or customization. A single motherboard can be customized to have different specifications, such as different amounts of memory. You might be able to upgrade the processor with a faster processor, and so on.

There are a number of other things on this motherboard, but one of the things that characterize it most significantly is its data rate, or its so-called *data bus*. The data bus is a description of the circuitry that carries data from the central processing unit, around the computer, particularly to the memory, as well as to any other circuits that are closely connected with the central processing unit. The speed of that data bus is really important, as is the width of the data bus. A data bus might be said to be 64 bits wide, or in a modern computer, it's more likely to be 128 bits wide. What that means is that when you send out data, 128 bits can travel down this data bus together, and go from memory to computer, or whatever. That would bring in two 64-bit words, or four 32-bit words, all at once.

Obviously, the wider that data bus is, the more information can be sent down it once, and it's also the speed of a data bus. A typical data bus can transfer hundreds of millions of bits of data every second, so that is a lot of data being transferred around with this computer.

There are also a number of specialized circuits on the motherboard. I don't know which these circuits are in this motherboard, but some of them are responsible for things like controlling, sending signals to and from the hard disk drive, for example, or to external circuits like a CD or DVD drive, or to the computer's monitor, which displays the information visually, or to just receive information from the keyboard, and so on. There are therefore a number of other specialized circuits that work intimately with the central processing unit, to begin to make what would be a whole computer.

In fact, you can think of a motherboard as basically being the entire computer, minus the peripheral devices that are nevertheless vital and let the computer be more useful to us and to communicate with others devices like the keyboard, the monitor, the hard disk that lets it store data permanently, when it's not even on, as well as the other peripheral devices you might choose to connect—a scanner, a digital camera while it's connected—any one of a number of things that you could connect to your computer. There are probably circuits on the motherboard that allow those things to be connected and that allow that to be a complete computer that's useful to you.

Let's take a look again at what a complete computer might look like. Here's a block diagram of a computer. A block diagram is an electronic circuit diagram in which you show different, major pieces of the computer, basically as just rectangular blocks. You don't try to show what's inside them.

We start out at the left here with the central processing unit, the CPU, the heart of the computer, the Pentium chip, the G5 chip, or whatever integrated circuit chip is that these days might hold one billion transistors, or something like that, and basically be responsible for all the processing that goes on inside of the computer.

Then there's the data bus, and I have represented this here as a sort of arrow with two heads, because the arrow shows that data is being transferred back and forth in two directions from the central processing unit to these circuits that handle peripheral devices, and from the peripheral devices back into the central processing unit.

For example, there's certainly random access memory, the subject of the previous lecture. That's a major additional part of the computer. It's usually right there on the motherboard with the CPU chip. The CPU chip needs to be able to communicate rapidly with the random access memory.

By the way, a comment about the sizes of computers that I should make: This information is flowing, ultimately, over copper wire, or wires made of other conducting materials, from, say, the memory to the computer. The speed of electrical signals in wire is typically on the order of about two-thirds the speed of light, so it's comparable to the speed of light, but not that high. It's pretty fast. The speed of light is 186,000 miles per second. You say, "Well, that's almost instantaneous." When you ask how long it takes light to go about a foot, the answer is about a nanosecond, about one billionth of a second. The speed of light is about a foot per nanosecond.

That means that computer signals travel about seven to eight inches per nanosecond, and that begins to have some significance for the building of fast computers. It means you cannot build a fast computer that is physically large, because if a computer is operating at 1GHz—which is a modest, relatively slow speed for a modern computer—that means that it executes individual operations in about one billionth of a second. If the memory were a foot from the central processing unit, it would take more than a second for the central processing unit to go out to memory, send it to an address, and then, in the another second or a little bit more, a nanosecond, or another billionth of a second, for the information to come back.

That would be very long, compared to the processing time of the computer. Today's computers, therefore—even supercomputers—have to be built small, so that the transit time for the electrical signals becomes insignificant compared to the basic clock rate of the computer. As clock rates go up, that requirement, set ultimately by the speed of light, becomes more and more stringent.

What else is on this computer? Well, other things are connected to the data bus. For example, there's a *disk controller*, which is the hardware that controls the hard disk, and then, there's the hard disk itself. Some of this controller is actually built onto the hard disk. I showed you a hard disk in the previous lecture; there was an electronic circuit board built right onto it, and that would connect with cables into one of these slots I have shown you on this board, so the hard disk would connect.

A couple of other things are connected: Keyboard and mice—we will talk about mice little bit later, and keyboards, and how they work. They are connected, and there needs to be devices called *serial to parallel converters*. I'll get to those in just a minute. There's the display, taking data off the data bus, being processed to display it in correct form on the display where other pictures or text are shown.

There's probably a USB port—*universal serial bus*. The universal serial bus is a wonderful mechanism that allows you to connect to any one of a broad number of devices from mice, to scanners, to digital cameras, to that computer. "Universal serial"—I'll get to "serial" in just a minute.

By the way, you might have noticed me, in the course of these lectures, kind of surreptitiously holding a little thing in my hand, and I once pointed out that it was a laser pointer, but that's not its main purpose. Its main purpose is as a little remote-control that advances the slides that I am actually displaying on the computer as I present these lectures. Since we usually hide that, because it's not part of the lecture, you don't see it going on, but in fact, since we're talking about computers, I've moved my computer a little bit, and of course, in a laptop computer, all of these things are combined in a nice, single case, very neatly done, with a flat screen, and so on.

I have connected to my universal serial bus here at the back, to a little receiver. That little receiver receives radio signals from this little clicking device, and every time I press the button, the slides advance by one on the computer screen. If I press the other button, they reverse back by one. And, the third button turns on the laser pointer. That's a simple little device that I've been using. And, the universal serial bus is sufficiently universal that it can accept the signals from that receiver that the computer manufacturer may not even have known existed when the computer was designed.

The universal serial bus works perfectly to make my computer do what I want it to when I press the button, so I press the button again, and down comes the optical drive control. That's connected to the mechanisms of the optical drive, and ultimately tells the laser when to fire if we are burning a CD, and it tells it when to move if we are reading information off of a CD.

That's kind of a block diagram of a complete computer. If I put that all together, the CPU, the random access memory, some of the disk controllers, some of these other converters are probably right on the motherboard, and then they connect by cables to peripheral devices that are located, perhaps, in different physical places, as with the external keyboard, or the external monitor of a desktop computer, or all integrated into the same case, as I have with my laptop computer here. That's what makes a complete computer.

Now, I mentioned universal serial bus, and I mentioned data bus, 128 bytes wide, and all of that. What's that all about? Well, there are really two ways to transmit digital information. There are *serial* ways of transmitting it, and *parallel* ways of transmitting it. I want to just give you a sense of the distinction between those two.

In parallel data transmission, you have a cable that can transmit lots of information simultaneously. Lots of bits go out on that cable simultaneously. There's a picture that shows a computer connected to anything—a printer, scanner, whatever—and if it's connected by a parallel cable, that cable has many, many, many wires. Those many, many wires are each simultaneously transmitting. In this case, there are four wires, and they're simultaneously transmitting four bits of information. Each of those bits could be a one or zero.

Parallel communication is fast, but it requires a big, ugly cable like this one. Look how thick it is. If you look at these connectors, and all the many, many, many little pins they have in there that have to be connected, that could get bent, and that could go wrong, it's a big, heavy, awkward cable.

However, with parallel data transmission, the data transmission is fast, because the data is also together. You could send a single word of data all at once on a cable with 64 wires, for example.

Much of our communication with peripheral devices these days is with serial cables. A serial cable needs only a single pair of wires, and with a serial cable, we send first one bit, and then a little while later, another bit, and then, a little while later, another bit, and then a little while later, another bit. It takes longer, but it takes a smaller, lighter cable. Examples of serial cable are this Ethernet cable, which would connect the Ethernet port of the computer to a high-speed network. It actually has four wires in it, two pairs. One pair is used for transmitting information from the computer, one pair is used to receive information, so that those two actions can go on simultaneously, and that makes it a little faster. In a sense, they've made transmit and receive in parallel with each other, but the data is sent serially.

Here's a USB cable, a universal serial bus cable. This end would go into my computer, and the other end would going to whatever device I'm talking about that I'm connecting it to, like my little receiver over there. USB cables also have two pairs, four wires. Two wires carry the data, and USB cables are also capable of providing a modest amount of electric power to peripheral devices, so that little receiver doesn't have to be simultaneously plugged into the wall. It doesn't need batteries, because the other pair of wires in the USB cable supplies a little bit of power. If you buy a USB hard drive, for example, it may be small enough, and use little enough power, that you don't need to simultaneously plug it into the wall. You simply plug it into the USB port, and the USB port supplies the power that's need.

That's the difference between "serial" and "parallel." If we go back to our block diagram for a minute, you can see that several of these devices—the keyboard and mouse, or the USB port—have serial to parallel, or parallel to serial conversion taking place, because the data bus is intrinsically parallel, but some of these devices are communicating serially.

By the way, I should point out that the one other very important part of the whole process I mentioned before—at the lower left—is the clock. It is, again, what determines that fundamental speed with the computer.

Let's talk a little bit about some of the other devices: Keyboards, for example, mice, and things like that. The keyboard is simply a set of switches. Here's a keyboard with a bunch of keys, and those keys are basically switches. This picture shows a sort of how they might work. Here are the keys, and under each set of keys—each row of keys, and each column of keys—there might be a wire, and when I press a key—for instance, the "F" key—that will join the two wires that cross at the "F" key, and electric current will flow in a circuit that comes out of that middle wire at the top of the microprocessor and goes into the middle wire on the side. That unique combination of wires carrying current is enough to signify that the "F" has been pressed. If I had pressed the "G" instead, the current would have been going in on that middle wire—on the left—but it would have been coming out in the right had wire on the top.

The microprocessor is increasingly built into the keyboard. The keyboard itself is a computer, and the microprocessor does several things. One thing it does it interpret what the key is, and then, it sends out the digital code. That happens to be the ASCII code for the letter "F."

Another thing it does—because mechanical switches can actually bounce a few times and send a few thousand "Fs" in the time that it takes to make contact—is that the microprocessor catches only the first contact and ignores the others for a while, for example. It has a lot of other functions like that. That's what microprocessors are doing. In some keyboards, there are actually what are called *capacitive switches*, where there's no actual contact made, and that solves some of those problems. Most involve actual physical contact, though.

The computer mouse is another important device. These are two fairly modern mice, because they're both wireless. They don't have to connect by a wire, but they send their information to a little receiver, similar to the one associated with my remote clicking device. The earlier mice had a rubber ball that actually rolled around, and there were optic or electromechanical devices that sensed the position of that ball.

More modern mice actually essentially have a tiny digital camera built in to them. In the bottom of the mouse, there's an LED, basically, which shines down on the surface, and there are an array of photo detectors very similar to, but cruder than, the detector in a digital camera. Basically, what happens is that the mouse takes a picture of the terrain it's rolling over, and as it moves, it's taking different pictures. By comparing them, it's figuring out where it's moving, and transmits that information to the computer, to tell it where the mouse is. A remarkable device.

We also need to have a display. Here's a modern display. It's a flat screen display, using liquid crystals. How do liquid crystals work? Well, I mentioned polarization in the last lecture, in the electromagnetic module. I talked about and showed you two polarizers, and showed how when you crossed them, no light could get through.

Well, liquid crystals are crystal structures that are nevertheless liquid, and the orientation of the crystal can alter the polarization of light. In a typical liquid crystal display, we have a grid that sort of sets up the orientation of the crystal in a certain way, and then, as you go through the display, the orientation of the crystal changes. That changes the polarization of light, so you have two polarizers, and the light comes through one polarizer. If it's had its direction changed by 90 degrees, it can't get through to the other polarizer.

However, when you apply an electric field to these crystals, they line up. There's an electric field. It's a positive charge on a metal plate at the top, and negative charge at the bottom. They all line up, and in that case, now, the polarization stays same, and the light can get through. What's happening in a liquid crystal display—here's picture showing the rather complicated structure of the liquid crystal display—is that ultimately, you're turning on and off individual pixels, individual little picture elements, these little spots on the screen, and for each different color of the three primary colors, you're doing that. You're doing that by applying little bits of electric field. In most crystal displays, these are thin film transistors; little transistors are actually handling the turning on and off of the electric field. That's letting light through, or not letting light through.

That's a quick look at how a computer and some of its peripherals work. In the remaining few minutes of this lecture, I would like to give you a little bit of a sense of how computers process what they do, because computers don't just process numbers. They process information—pictures, music, and text. Let's take a look at how some of that works.

I've showed you some of the ASCII codes before, for coding text information. Here's a table that has some. Character A. The decimal is 65. In binary, it's 0100001 at the right. It's also listed in two other funny unit systems, hexadecimal, and octal. I will tell you in a minute why those are there. Here's a lowercase "p;" that was capital "A."

Here's the space. It's decimal 32. It's hexadecimal 20. It's octal 40. It's binary 01, with the rest zeros.

There's the letter "N." It's 78 in decimal. It's 4E in hexadecimal. It's 116 in octal, and it's that long string in binary. What's that all about?

Well, let me tell you a little bit about numbers. You may have learned this sometime way back when you learned new math in eighth grade, or something. That was called the "new math," in my day. It's certainly not new now.

Here, the digits to make the base 10 numbers that we always work with: 0, 1, 2, 3, 4, 5, 6, 7, 8, and 9. Out of those, we build all the possible decimal numbers.

Here the numbers in base two. Simpler: 0, 1. The trade-off for them being simpler is that we get longer numbers.

Here are the digits in base eight. They are: 0, 1, 2, 3, 4, 5, 6, 7, and that's it; so, we get slightly longer numbers in base eight. A one in the second place means one times eight, not ten.

Here's hexadecimal: 0, 1, 2, 3, 4, 5, 6, 7, 8, 9, A, B, C, D, E, F, because there are 16 possible digits in hexadecimal. The second decimal place is 16, and the third is 16 squared or 256.

The reason we like octal and hexadecimal is because those are powers of two. Eight is two cubed, 16 is two to the fourth power, so that it's very easy to convert from octal and hexadecimal to the true binary numbers that are part of the computer, and that's not so easy for base 10.

To give you an example, N is 78 in base 10. That's 7 times 10 plus 8 times 1. That's 78. N is 116 in octal. That's 1 times 8 squared, which is 1 times 64, plus 1 times 8, plus 6 times 1, and that is 78.

Here's N in hexadecimal. It's 4E16. Well, that's 4 times 16, because that's 16th place—this is only a two-digit number in hexadecimal – 1 times E—E is—F is 15, if you look above at that row, in hexadecimal. E is there for 14. That's 1 times 14, and that comes out to 78 as well.

If I convert that to binary, all I need to do to that hexadecimal number is take and convert each of the digits. The first one is a 4, so that's 0010 in binary, and the second is a 14, so that's 1110, and that does that conversion.

Those are the number codes that correspond to letters. What about when we're dealing with not text, but a picture? Well, digital cameras take an enormous amount of information. Let's take a look at a typical picture. Here's a photograph of my dog Fergus. He's a Shetland sheep dog, and he's romping in a snowstorm. I took a picture of him with my four-megapixel digital camera. That's four million pixels. Each of those pixels has 24-bits, which represent the intensity of the light, and the color. There are 2 to the 8 levels of intensity. That takes eight bits, in each of three colors. If you put that altogether, that's about 96 million bits. A bit is eight bytes, so that's about 12MB.

The picture file—and this would be true if I downloaded it in its raw format from my camera—would occupy about 12MB of space in my computer, and would quickly fill up my computer's memory and hard disk with pictures.

I don't need that much information. I can contain almost exactly the same information with far fewer bits. Here's how it works. We have some kind of compression scheme, so let's blow up this part of Fergus where his white ruff joins his browner fur. Let's blow that up again, and let's blow that up once more. You can see the jaggedness in that circular blow up, and in this next blow up, you can actually see it divided into 25 little squares, each of which is identical in color and intensity over that square. These are the individual pixels. There are 25 of them in this little, tiny place.

How much information is there? Well, 25 pixels. There are 24 bits per pixel, so that has 600 bits of information. There are many schemes for compressing that information into less, and I'm just going to give you an example. That's related to, but not the same as, the so-called JPEG compression—Joint Photographic Experts Group—, which is a common scheme for compressing photographs. It works something like this.

The compression scheme says, "Let's start at the left side of this, because the color is pretty uniform along the left side." Remember that this is in the region where the color is changing from white to brown. It's pretty uniform along the left, so that it takes two bits to specify where we are—left side, right side, topside, bottom side—four possible sides. Two bits, four possible choices, so that it takes only two bits to specify that.

We are going to start at the left side. The color and the intensity on the left side takes 24 bits, so we have two bits to show where we're starting, and 24 bits to describe what's going on there. Then, let's say move to the right, gradually changing uniformly to the right edge color. Well, it takes 24 bits. We already know since we're on the left that we're going to progress to the right. That doesn't take any more information. We need 24 bits to talk about the intensity and color at the right edge. That's 24 more bits. If you add it all up, we have the two bits for the left edge starting point, 24 bits for the colors and intensities at those two edges, and we know how to fill in all the rest, because a computer can compute. We have reduced the total bits to 50, a compression of about 12 to 1.

That's actually fairly typical of the compression schemes that are used in digital cameras. There are other compression schemes used to get movies onto DVDs. They have to be very clever. They generally work well if you have gradual transitions. They tend to fail where you have very abrupt transitions. Sometimes you'll see funny little graininess jumping into a digital movie. That's why it's occurring. The compression scheme is not perfect. That's how we process massive amounts of data like pictures.

Let's now wrap up this entire module, and see where we have come from. Again, we started at silicon atoms. In the previous module, we got all the way to things like memories. In the module before that, we got all the way processing circuits. Now, we've seen the complete computer. We've put that together. We've added peripheral devices, like LCD displays, liquid crystal displays, keyboards, mice, and those kinds of things. We have seen how we store everything in the form of binary numbers, and how those binary numbers might represent particular decimal numbers, they might represent letters, they might represent digitized parts of digital images, and in the end, we get something we're interested in, like a true digital picture. That's all the way from the atom to the computer.

Module Five: Fire and Ice
Lecture Twenty-Five
Keeping Warm

Scope: Phenomena involving heat play a major role in our lives. We humans can survive only in a narrow temperature range. The Sun's energy keeps Earth's temperature near that range, while technology provides further control over our indoor thermal environment. Technically speaking, *heat* refers to energy that's moving from one place to another as a result of a temperature difference. Both natural and technological systems exploit a variety of mechanisms to facilitate such heat transfer. Understanding heat transfer lets us design more efficient buildings, make instantaneous measurements of body temperature, diagnose disease, and cook delicious food. On a larger scale, the same principles let us take the temperature of the stars and of the Universe itself. And understanding heat transfer helps us recognize the changes we humans are causing in Earth's climate.

Outline

I. *Heat*, loosely, refers to the energy associated with random motions of the atoms and molecules that make up matter. Strictly speaking, that energy should be called *internal energy* or *thermal energy*. *Temperature* is essentially a measure of thermal energy on a per-molecule basis. The term *heat* describes a flow of energy resulting from a temperature difference.

   A. Common temperature scales, including degrees Fahrenheit and degrees Celsius, are based on everyday occurrences, such as the freezing and boiling of water. However, because temperature is a measure of energy, there's an absolute minimum temperature, corresponding to the lowest possible energy of a system. This *absolute zero* lies about 460 degrees below 0 Fahrenheit and 273 degrees below 0 Celsius.

   B. Temperature alone tells nothing about how much thermal energy an object contains. We're usually less interested in total energy in an object than in the energy associated with a change in its temperature. The energy involved per degree of an object's temperature change is the object's *heat capacity*.

1. For equal amounts of material, different substances have different heat capacities.

2. Water has a particularly high heat capacity. That's why it takes so long to bring water to a boil on the stove, and it's also why large lakes have a moderating effect on the surrounding climate.

II. When an object's temperature differs from its surroundings, energy flows from the object to the surroundings (if the object is hotter) or from the surroundings to the object (if it's cooler). Because this energy is flowing as a result of a temperature difference, it's properly called *heat*. There are three common mechanisms for such heat transfer:

A. *Conduction* occurs when two objects or substances are in direct contact. Particles in the hotter substance are moving faster, and when they collide with the slower moving molecules in the cooler substance, they transfer energy. Because the rate of heat transfer is proportional to the temperature difference, a greater temperature difference makes for a greater heat flow. That's why it takes more energy to heat your house in colder weather.

1. Materials differ in their conduction effectiveness. Metals are very good conductors of heat, while water is moderately good and air is a poor conductor. Such materials as Styrofoam and fiberglass insulation, which contain trapped air with very little solid material, are designed to be especially poor heat conductors and, thus, to inhibit heat flow.

2. The thermal conductivity of building materials is what ultimately determines how much energy we must supply to heat or cool our houses. The familiar R-value used to rate insulation is a measure of insulation effectiveness. R-19 fiberglass insulation, for example, loses 1/19 of a British thermal unit of energy for each square foot of area, for each degree Fahrenheit temperature difference from one side of the insulation to the other. (One Btu is the energy needed to raise a pound of water 1 degree Fahrenheit; it's a unit used almost solely in the United States and is analogous to the calorie, which is the energy needed to raise 1 gram of water 1 degree Celsius.) Because R-values add, it's easy to figure out the heat loss through a composite structure, such as a wall.

3. I'm often asked whether it makes sense to turn one's heat down at night, because of the need for extra heat to warm the house in the morning. Pause a minute and think about this, given what you know about heat transfer by conduction. In fact, you're always better off turning down the heat, because a lower temperature means a lower rate of heat transfer.

B. *Convection* is the transfer of energy by the motion of a fluid, such as air or water. Because warmer fluid is less dense, it rises. As it does so, it gives up heat to its cooler surroundings. The now-cooled air eventually sinks, resulting in a continuous circulation of fluid that transfers heat from a warmer, lower level to a higher, cooler level.

1. Convection is responsible for several everyday heat-transfer occurrences. When you heat water on the stove, heat flows from the stove into the water by conduction. Heating at the bottom of the water then sets up convection currents that transfer heat throughout the water. On a larger scale, heat sources in a house set up gentle convection currents that circulate warm air throughout the house.

2. Convection is also important in natural systems. Giant convection cells associated with strong heating in Earth's tropics help transfer energy poleward, making the planet a more uniformly warm place than it would otherwise be. Coupled with Earth's rotation, these convective flows also result in the prevailing west-to-east winds in Earth's mid-latitudes. Convection just below the visible surface of the Sun transports the enormous energy generated in the Sun's interior outward to the surface. And Earth's continents drift about slowly on convection currents resulting from rising heat in Earth's mantle. Finally, convection in the electrically conducting fluids in the interiors of Sun and Earth are responsible for these bodies' magnetic fields.

3. It's because of convection that an air space alone is not very effective at preventing heat loss in buildings. Properly designed insulation, such as fiberglass, Styrofoam, or goose down, traps air and keeps it from moving, thus exploiting air's poor thermal conduction while preventing energy loss by convection.

C. *Radiation* is the transfer of energy by electromagnetic waves. At temperatures above absolute zero, all objects emit electromagnetic waves, because the thermal motion of their constituent particles involves the acceleration of electric charge (Lecture Eighteen). When an object is warmer than its surroundings, radiation results in a net loss of energy. Radiation increases dramatically with increasing temperature (as *temperature*^{4}), so it's particularly important at high temperatures. A hot stove burner and the filament of a lightbulb are both losing energy almost entirely by radiation; so is the Sun itself and, for that matter, so is the Earth. When an object—such as a planet or star— is surrounded by vacuum, then radiation is the only means of heat transfer available.

    1. Shiny objects reflect radiation, which is why building insulation is often covered with aluminum foil. That's also why vacuum ("thermos") bottles are shiny on the inside; the vacuum prevents heat flow by conduction or convection, and the shiny coating turns back radiation. This is also why high-quality windows have so-called low-E coatings that inhibit energy loss by radiation.

    2. Not only does the amount of radiation increase with temperature, but so does the frequency of the waves emitted— and frequency is related to color. Thus, it's the temperature that ultimately determines the color of a hot, glowing object. The Sun, at nearly 6000 °C, glows with visible light, essentially white. A lightbulb filament is at about 3000 °C, and it glows with a yellower light. Turn down the current to the filament, and it gets redder. Even when an object is too cool for its glow to be visible, it still radiates—now in the infrared or even radio region of the electromagnetic spectrum.

III. With no additional energy, a hot object will gradually lose energy by heat transfer to its surroundings, eventually reaching the same temperature as the surroundings.

    A. To keep an object warmer than its surroundings requires a supply of energy. If the energy supplied to the object exactly balances the energy lost by conduction, convection, and/or radiation, then the object's temperature will remain constant. This is the state of *thermal energy balance.*

        1. In buildings and other engineered systems, a temperature-sensitive switch called a *thermostat* is used to turn the energy source on if the temperature drops below a set point and off when it rises much above that point. The direct effect is to maintain the desired temperature; consequently, the energy supplied to the building is automatically made equal to the energy lost.

2. In warm-blooded mammals, including humans, the brain's hypothalamus monitors changes in body temperature and orders a variety of compensating mechanisms to maintain a constant temperature. These include changes in metabolic rate, changes in the configuration of hair or fur, sweating, and constriction or dilation of blood vessels. The result in all cases is to maintain internal energy generation at a rate that balances energy loss to the environment.

B. Even without active regulation mechanisms, an object with a fixed energy supply will automatically reach a state of energy balance. That's because all three heat-transfer mechanisms increase the rate of heat flow as the temperature difference increases.

1. Consider, for example, a solar-heated greenhouse with no other energy source. If it's at the same temperature as its surroundings, then there's no heat loss. However, because there's energy coming in from the Sun, the greenhouse temperature goes up. But then heat begins to flow out. As long as the rate of outflow is less than the rate of solar energy coming in, the greenhouse will continue to warm. As its temperature increases, however, so does the heat loss. Eventually, a temperature is reached at which the heat loss exactly balances the solar input. That's the energy-balance temperature, and as long as the solar input stays constant, the greenhouse will maintain that temperature. If the solar input drops, the greenhouse will lose heat at a greater rate than it gains solar energy, so it will cool down. But then the heat-loss rate drops, and eventually, a new balance is established.

2. This phenomenon of energy balance is what determines temperatures in a wide variety of technological and natural systems. Why is the filament of a lightbulb at about 3000 °C? Because at that temperature, the rate of energy loss by radiation is equal to the rate at which electrical energy is supplied to the filament. If it were cooler, the energy loss would be lower and the electrical energy input would exceed

the loss; the filament would heat up. If it were hotter, the loss would exceed the electrical input, and it would cool down. Thus, a state of energy balance is automatic. Why is the Sun at 6000 °C? Because at that temperature, the rate at which it loses energy by radiating sunlight balances the nuclear energy generation in the Sun's core.

3. Energy balance is the ultimate determinant of Earth's climate. We'll take a closer look in the next lecture.

Suggested Reading:

Louis A. Bloomfield, *How Things Work: The Physics of Everyday Life*, chapter 6.

W. Thomas Griffith, *The Physics of Everyday Phenomena: A Conceptual Introduction*, chapter 10.

Paul Hewitt, *Conceptual Physics*, chapter 16.

Going Deeper:

Richard Wolfson and Jay M. Pasachoff, *Physics for Scientists and Engineers*, chapter 19.

Questions to Consider:

1. The temperature of a candle flame is about 2500 °F. A bathtub full of water might be at 100 °F. Which has more "heat" or, to be precise, more thermal energy?

2. What will doubling the R-value of all the walls and windows in your house do to your heating bill?

Lecture Twenty-Five—Transcript
Keeping Warm

Lecture Twenty-Five, "Keeping Warm." This is the first lecture in Module Five. The title is, "Fire and Ice." As the title of Module Five, "Fire and Ice," and the title of this lecture, "Keeping Warm" both implies, this module is about heat and related phenomena. Let's take a look, as we have with the other modules, at a quick outline what we're going to be doing in the six lectures that comprise this module.

The first lecture is called, "Keeping Warm." I'll say a bit more about what it's going to do in just a minute. Lecture Twenty-Six, "Life in the Greenhouse" is going to talk particularly about Earth's climate, about how this climate is established, and about how we human beings may be changing Earth's climate.

Lecture Twenty-Seven, "The Tip of the Iceberg" talks about changes in state between liquids, solid, gases, and how they affect different substances. It will answer question I hinted at before in my lecture on taking flight, of all things: Why the iceberg floats with most of its matter submerged; in fact, why the iceberg floats at all, which is a remarkable question. We will get the changes in state of materials with changes in temperature, and generally, thermal conditions.

"Physics in the Kitchen." I love to cook. The kitchen is full of examples of physics: Heat transfer occurs, we cook things with microwave ovens, we need to know how to get an egg just right, we need to explore different cooking techniques—how is baking different from broiling, different from boiling, different from steaming. I love double boilers because they cook custards, particularly tapioca, to perfection. How does that work? We will look at physics in the kitchen.

Lecture Twenty-Nine is, "Like a Work of Shakespeare." Am I going to talk about the thermodynamics of science? Am I going to talk about Hamlet's temperature? What am I going to talk about there? "Like a Work of Shakespeare" is a quote from the British writer C.P. Snow, and I will leave it at that until we get to Lecture Twenty-Nine.

Finally, Lecture Thirty, "Energy in Your Life." We will talk about ourselves as creatures in 21st-century industrialized society, and the amount of energy it takes to supply us with our energy needs. We got a hint of that several lectures ago when we had somebody cranking—it was Jaimee—away on that hand crank generator. We will bring that all up again, and we will talk about how much energy is actually used in your name and where that energy comes from. As that last lecture's title hints, energy and thermal phenomena, temperature and heat, are closely related.

Well, this is Lecture Twenty-Five, "Keeping Warm," and as the title suggests, we are going to be talking about how it is that systems keep at the particular temperatures they're at. We human beings, in particular, are able to exist in only a relatively narrow range of temperatures. We can survive only over a few tens of degrees Fahrenheit, and we are comfortable only over a very narrow range of a few degrees to 10 degrees or so Fahrenheit.

We are very restricted as to our comfortable temperature ranges. How do we manage to keep warm? How we keep our buildings warm? How does our planet keep warm? How does the greenhouse keep warm? How do natural systems keep warm? What determines the temperature of the Sun? What determines the temperature of the filament of a light bulb? How do those things keep warm? Those are the kinds of questions we're answering this lecture.

Let me begin with some definitions. We have sort of talked loosely about heat. I want to be more precise about the definition of "heat" then you probably are in your everyday life. We sort of think of heat—and I have suggested this before—as referring to random motions of the molecules that make up a substance. I argued before that for example, when I do something like step on the brakes in my car—providing that it isn't a hybrid car with regenerative braking, using the law of induction, and putting that energy back into the battery—if it's an ordinary car with friction brakes, what we are doing is turning the directed forward motion of the car into this random motion of the molecules that we call, loosely, "heat." I have always used that word that way, because it isn't quite the definition of "heat."

Strictly speaking, that energy associated with the random motion of molecules should be called *internal energy*. It is energy that is internal to a system. It isn't associated with the bulk motion of a system. It is internal to the system in the sense that it belongs to the individual particles, the individual atoms and molecules, which make up that system. What we tend to think of as heat—I said, "There's a bathtub full of hot water. It's got a lot of heat in it." I really shouldn't say that. It's got a lot of thermal energy.

What is heat, then? "Heat" has a very specific meaning in physics. It means energy that is in transition, going from one place to another. And, not just any energy that is transition, but—for example, the energy that is going down an electrical wire to bring power to my laptop computer right now isn't heat—heat is energy that is flowing because there is a temperature difference between two things, or two places. When there's a temperature difference between two things and two places, energy tends to flow from one to the other, and the energy flowing as a result of the temperature difference is called "heat." Once the energy gets into the other substance, it is then, strictly speaking, internal energy.

You may say, "Why are you quibbling with words like this?" The answer is an important one. If I have a bathtub of cold water and I make it hot, one way of making it hot would be to put it in contact with a hot object. I could dump a hot block of stone into it, heat up a stone in fire, and dump it into the bathtub, and the bathtub would warm up because the stone would be hotter than the bath water and energy would flow as result of the temperature difference.

I could equally take that bathtub, and put my hand in it, and agitate it vigorously, supplying energy, and turning that energy by friction into the internal energy. My hand is not significantly hotter than the bathtub, and may be even cooler. It's not transferring energy by virtue of a temperature difference, or by another means, but the end result is the same. The bathtub is at a higher temperature, and it has more internal energy. "Heat" refers to energy that is in transition, specifically because of a temperature difference.

Now, I have talked about temperature differences. What on earth is *temperature*? Well, temperature is actually a fairly simple thing, in the easiest physics definition. It gets pretty complicated in more complicated definitions. In the simplest definition, temperature is simply a measure of the average energy, the average internal energy that individual molecules have. You have to be a little bit careful with that definition. It's most accurate for simple dilute gases like the atmosphere around you, but in basically place. What temperature is basically measuring is how much of that internal jiggling energy there is on average. If you increase that energy, you have increased the temperature.

Now, there's a whole range of internal energy, the internal speeds of the particles racing around in this room at speeds on the order of 300 meters per second. Some of them are going 1,000 meters per second, and some of them are going 10 meters per second. The average speed though, is around 300. That's also close to the speed that you're most likely to find.

Temperature is a measure of the energy, ultimately the kinetic energy of the molecules as they whiz around. We have heat, therefore, which describes how much total energy, internal energy, energy associated with the random motions in something—I'm sorry. Not heat. "Heat" is the energy flowing. Internal energy is that total amount of energy in a substance. Heat is energy that flows from one thing to another as a result of a temperature difference, and "temperature" is a measure of the average amount of energy in the molecules in a substance, and the average amount of internal energy per molecule. They're all measuring slightly different things.

How do we measure temperature? Well, temperature unfortunately, like the meter-kilogram system of units that we invented to do electricity and magnetism is really—the temperature scales that we have really aren't quite as natural as they might be. Because temperature is a measure of energy, and because systems have a minimum of energy they can possibly have—you might think that minimum is zero, but because of quantum physics, it isn't quite zero; it's tiny, though—there's actually a minimum temperature that a system can have, and that temperature is called *absolute zero*.

If we had been smart, and physicists who worked with thermal effects particularly had been smart enough to use the system of temperature, we would use a temperature system based on absolute zero. That's not temperature system we use. In the United States, we tend to use the Fahrenheit scale for temperature. I have here a picture here that shows as horizontal lines the temperatures at which various phenomena occur.

At the very bottom is absolute zero. You cannot go lower than that. It is like the bottom of this swimming pool. You can't have any less water than if you were to take it all out. There's a temperature close to absolute zero at which nitrogen boils. In later lectures in this module, I'm going to be working with liquid nitrogen. It's going to be at about that temperature. Much higher than that is the temperature at which ice melts. Higher still is the temperature at which water boils.

On our Fahrenheit scale, 212 degrees is the boiling point of water. Thirty-two is the freezing point of water, the melting point of ice. On that scale, nitrogen happens to boil at about -321 degrees Fahrenheit. Absolute zero occurs at about -460 degrees Fahrenheit. Most other countries besides the United States, and most scientists who don't have to use the absolute zero scale—for instance, a biologist talking about the best temperature to incubate these cells—would typically use the Celsius scale.

The Celsius scale is quite similar to the Fahrenheit scale in that it doesn't have its zero at absolute zero. On either of these scales, "zero degrees" doesn't mean, "zero" of something. It's just an arbitrary point on the scale. At my home in Vermont last winter, it got to 20 below zero many times. There isn't anything magical about that negative temperature. We're still far above the true absolute zero. That's just an arbitrary point on an arbitrary scale.

The Celsius scale is a little bit more rational in that the melting point of ice, the freezing point of water, is zero degrees Celsius. The boiling point is 100 degrees higher at 100 degrees Celsius. Nitrogen happens to boil or liquefy—depending on which way you are going—at almost -200 Celsius, and absolute zero is -273.

Neither of these scales is the best to use. The actual official international system of units temperature scale is the Kelvin scale. On the Kelvin scale, absolute zero is zero Kelvins. By the way, if you want to sound very sophisticated, say "degrees Celsius," "degrees Fahrenheit," but Kelvins are an official international system unit, and you just say "Kelvins." If you hear a physicist say, "Ah, the Sun is at 6,000 degrees Kelvin," you can say, "You aren't quite using the right terminology. The Sun is at 6,000 Kelvins."

Ice melts at 273 Kelvins, and water boils at 373 Kelvins, 100 Kelvins higher, or 100 degrees Celsius. Nitrogen boils, or liquefies, at 77 Kelvins, and absolute zero is, as it should be, zero Kelvins. Notice that there are 100 degrees between the boiling point of water, down to the melting point of ice on the Celsius scale, and that there are 100 Kelvins from the boiling point of water down to the melting point of ice. That same 100 difference means that a degree of Celsius change is the same as a Kelvin change. A Kelvin, and a degree Celsius, have the same size. The only difference between those two scales is the zero point. The Fahrenheit scale is quite different. The degree size is different, and the zero point is also different.

There's a fourth scale I'll just mention. Unless you are an engineer, you've probably never heard of the scale, and have had no reason to. Engineers who deal with things like power plants need to work in absolute temperatures, and so, engineers who work in the United States, at least, use the *Rankine scale*, in which a degree ranking is the same as a degree Fahrenheit in size, but the zero point has been put correctly at absolute zero. On that scale, ice melts at 492 degrees Rankine, and water boils at 672 degrees Rankine. We'll see in subsequent lectures, particularly that one about Shakespeare, just why we need the temperature scales that begin at zero, particularly when we're talking about things like power plants, and things like that.

Now, should mention that we are rarely interested in the absolute amount of energy in some warm substances. For instance, let's go back again to my bathtub of warm water. I don't really care how much total energy is in there. The Shakespeare lecture is going to give us a better reason as to why that is, but one reason is that it turns out to be impossible to get all that energy out, even if you would like to.

What we're usually more concerned with his how much energy it takes to change the temperature of a substance or an object by a given amount. That number is called the object's *heat capacity*. The bigger the heat capacity, the more energy the object can store without changing its temperature very much. Water happens to be a substance with an enormous heat capacity when compared with most other substances, or very large, anyway. That's why it takes water a very long time to boil when you put it on the stove. It takes quite a bit of energy to raise the temperature of water from tap water temperature up to that boiling point of 212 degrees Fahrenheit, or 100 degrees Celsius. Takes a long time to do that, because it takes a lot of energy coming in from your stove to do that.

It's also why large lakes, like the Great Lakes, or the ocean, exert moderating effects on the climate around them, because it takes a long time for those temperatures to change, so they act like temperature stabilizing regions. Regions near large bodies of water tend to experience less extreme temperature changes, at least as long as those bodies of water are not frozen.

Now, I've indicated that heat is the flow of energy resulting from a temperature difference. Therefore, when an object's temperature differs from its surroundings, heat will flow either into the object if it's cooler than its surroundings, or away from the object if it's warmer than its surroundings, and that is properly called "heat." There are three mechanisms whereby energy is transferred by heat, or, to put it more succinctly, three different heat transfer mechanisms, which are important to understand if we are to answer the question that is implied in the title of this lecture, namely, how do we keep things warm?

The most common mechanism is *conduction*. It occurs when you simply place two objects in direct contact with each other. What happens is that in the hotter object, particles are moving a little faster. They collide with slower moving particles in the attached cooler object, and they transfer their energy. Think back to the last lecture in the module on motion, Module

Two. I showed you some collisions of two parts, and I put different masses on them. One object was moving faster, and hitting a slower moving object, and energy got transferred to it. I similarly bounced a couple of balls—a tennis ball off of a kickball. That kind of collision occurs at the microscopic level; and in the objects are moving faster, the molecules in the hotter substance, energy is transferred to the cooler substance. That is the mechanism of heat transfer.

Different substances, again, are better conductors of heat, better at causing this heat conduction to take place. Metals happen to be very good conductors of heat, and it is for the same reason, basically, that they're good conductors of electricity. They have a lot of free electrons. Those free electrons are lightweight. They easily respond to collisions, or a warmer region of a piece of metal, and they quickly move and transfer that energy to other parts of the sources. Thus, a metal, if you heat up one part of it, quickly becomes hot all over. If you stick a hot poker in the fire and leave it there very long, you can't even touch the metal parts of it that are far out of the fire, because metals are very good heat conductors for the same reason they are good electric conductors.

Water is a moderately good conductor. Air, with its widely separated molecules, tends to be a relatively poor conductor of heat. Materials like Styrofoam or fiberglass insulation manage to trap heat, and I have a piece of Styrofoam over here that's been cut open. You can see the little bubble-like structure in it. They happen to trap heat, because they stop heat flow. They have little trapped air bubbles, and air is a relatively good insulator, so that basically, heat doesn't move through Styrofoam because of those little bubbles of air, which are very poor conductors of heat.

Now, I usually like to avoid the non-metric units. You notice that I've been trying to talk about meters here, and I would talk about Celsius degrees and things like that, because those are the units that scientists use. I do want to say something about one at that's worth knowing about. That is the so-called "R-factor," used in measuring insulation effectiveness.

Over here, I have a piece of Styrofoam blue board. This is commonly used to insulate basements, for example. It's applied to the outside the basement walls, before a new home is backfilled. Styrofoam has the property that has an R-value, which I have designated by this script "R" here, of six [sic five] per inch. Now, there, R-value has meaning. This is 1.5-inch thick Styrofoam, so it has an R-value of 7.5 total. That R-value has meaning, or rather, the inverse of that R-value has meaning. One over that R-value has a specific meaning in terms of English units. Here's what it means. One over 7.5, if you do the math—arithmetic—is about 0.13, a little more than one tenth.

The units of that, which we rarely give—if you go to your construction yard and buy some insulation, and ask, "What's the R-value?" They'll say, "This is six-inch fiberglass: R-19." If you ask them what the units are, they probably won't be able to tell you, but it does have units. The units are British thermal units. That's a unit of energy. It's the amount of energy that it takes to raise the temperature of a pound of water by one degree Fahrenheit. It is analogous to the calorie, which is the amount it takes to raise a gram of water one degree Celsius. The British thermal unit—used in the United States, not in Britain—per hour, per degree temperature difference, between the two sides of the Styrofoam, per square foot. What that means is that if I have one square foot of this Styrofoam insulation, and the temperature difference between the inside and the outside is one degree Fahrenheit, then every hour, this will lose 1.3 BTUs of energy.

If I have many, many square feet in my walls, I multiply by the number of square feet. I have. If it's 68 degrees Fahrenheit inside my house, and it is zero degrees Fahrenheit outside in the winter, I multiply by 68 degrees for that difference.

That number it is what allows, say, the plumber or heating specialist who is sizing a new furnace for your new house to calculate how much heat output that furnace has to be capable of, because they can calculate from the R-values of your insulation what the energy loss will be. That's what that R-value and insulation is all about, and in fact, R-value simply adds to very useful numbers to deal with.

By the way, people are often asking me, "Should I turn down my thermostat at night? Does it make sense to do that, or does it does take more energy to heat up in the morning?" The answer is obvious if you think about it, so think about it. Okay? Why is the answer obvious? Because the rate of heat loss depends on the temperature difference. So, if my house cools down at night, the heat loss rate is going to be lower, and I'm not going to have to over compensate for that, to bring the heat up in the morning. That's conduction, and that's one of the important things in determining energy losses from our houses.

Another heat loss mechanism is *convection*. "Convection" is the transfer of energy by the bulk motion of a fluid, and it's responsible for number of common energy transfer mechanisms. For example, heat flow by convection—here's a symbolic picture of a pot of water on the stove. The stove burner is red-hot. There's heat transfer by conduction into the water, but then, the warmed water becomes less dense in the middle, it rises, sinks at the edges, and sets up these patterns of circulation—they're called convection cells, and are ultimately what transfer energy from the bottom of the water to the top of the water.

That's a very common convective heat flow. Convective heat flow—you have a radiator in your house that heats up the air, and that sets up a convective air circulation, transferring heat around the room, for example.

Convection is also important in a number of natural systems. Here's a picture on the left, looking down on convection cells in the laboratory, and they form these were remarkably distinct hexagonal patterns; in the middle of these cells is where the warm fluid is rising. Those hexagonal boundaries are where the cool fluid is falling, having transferred its energy to whatever is that the top of this picture.

Here's a picture of the Sun, taken through a special telescope. The Sun has convection going on, and it transfers energy from its interior to its surface—this is actually a movie. If I get the movie started, you can actually see those convections. This is actually the boiling motion occurring on the surface of the Sun. Let's run that again. There's the surface of the Sun, the turning common convective motions as the convection occurs inside the Sun and brings that "boiling" fluid to the top. It's not boiling, but it's moving in these convective motions. It's not a liquid. It's a gas, in the case of the Sun.

A much slower version of natural convection occurs inside the Earth. Below the crust of the Earth is the semi-plastic-ish mantel. You can think about it as a very, very thick viscous fluid. There's heat rising from the interior of the Earth, from the primordial heat, from when the Earth was formed, and also from radioactive decay. That sets up convection currents in the mantel. They run very slowly, on time scales of tens of millions to hundreds of millions of years. That's what drives continental drift. For instance, the North Atlantic Ocean is opening, fundamentally, due to convection. Patterns like this are carrying Europe and America apart at about the rate that your fingernails grow. It's measurable. We can actually measure it with things such as the global positioning system, which I will talk about in a later lecture.

By the way, I mentioned that air is a poor conductor of heat, and therefore, a good insulator, but it isn't good enough to just make, for instance, a double-paned window, with a gap between the two windows, because if you have air, air readily sets up convection currents. Those convection currents can carry heat, even if air isn't a very good conductor. That is, again, why we use insulation like fiberglass and Styrofoam, because those little bubbles inhibit the convective motions of the air. Then, we get the true insulating value of the air, because we've suppressed the convection.

Finally, let me mention an energy loss mechanism that here on Earth tends only to be important at very high temperatures. It's the way a hot wood stove sends much of its energy into the room, for example, or a light bulb filament—that's radiation. That's the transfer energy by electromagnetic waves.

Because all objects involve vibrating molecules and atoms, and they have electric charges, they represent accelerated electric charges. As we saw in the last lecture of Module Three, electric charges are what produce electromagnetic radiation. Something like a hot stove burner then, emits electromagnetic radiation that becomes visible for you to see.

Shiny objects reflect radiation; so shiny objects can block the flow of radiation. Here's a thermos bottle. If you look inside a thermos vacuum bottle, you see that is shiny on the inside. That is to suppress any energy loss from radiation. In addition, there's a vacuum between the inner and outer walls of these containers. Conduction and convection can't occur in a vacuum, because there are no materials to handle the conduction or convection, so the thermos bottle effectively suppresses virtually all energy loss from what you put in. That's why it's so smart that you can put cold stuff in it, and it stays cold, or hot stuff in and it stays hot. That's also why some very high-quality windows that you might buy for a new house are called "low-E windows." They actually block the flow of infrared radiation that your house would tend to emit.

The hotter an object gets, the more electromagnetic radiation it emits, and something else happens as well. That something else is that the color of the radiation changes. Here I have an example of that. Here I have a light bulb with a bare filament. You can see it. It isn't a frosted light bulb, like most light bulbs you use. I have it connected to this device, which is actually a variable autotransformer. It uses, again, Faraday's law of induction, to provide a variable voltage to this light bulb. As I turn this light bulb up, at first, not much happens. However, there is current flowing through it. You just don't see much happening.

As I turn it up though, it begins to glow, at first a dull red. It's not emitting much radiation, and the radiation is emitting is primarily not that red color, but is actually infrared, which I can't see. As I turn it up more, two things happen. It emits more radiation—and by the way, the amount of radiation it emits goes up very rapidly, to the fourth power of the absolute temperature—so as I double the temperature in Kelvins or Rankine, not from 10 to 20 Fahrenheit, but from 100 to 200, or from 200 to 400 Kelvins, I will multiply the amount of radiation by two to the fourth, or by 16 fold.

It therefore goes up very rapidly with increasing temperature; and furthermore, you'll see that the light is getting more and more yellow, and more and more of it is coming out now. It's visible light, more and more visible light. More and more light, and it is getting whiter, and whiter. I'm actually exceeding the designed voltage of this lamp, so it may not last long. There it is, white-hot light. The temperature of that filament is reflected in the color of the light it emits.

The Sun, for example, is white-hot. It is at 6,000 degrees Celsius, about the same as 6,000 Kelvins, because there is only a few hundred degrees' difference in the zero point. The filament of a typical light bulb—not when I had that all cranked up—is about 3,000 Kelvins, about 3,000 degrees Celsius. I had that cranked all the way up. It was almost same temperature as the surface of the Sun, probably. I can tell that because it was glowing at about the same kind of white temperature.

I want to end with an important concept, which we will use more of in the next chapter. If I have a hot object, and I add no additional energy to it—for example, this light bulb—and I turn off the power, so that there's no energy source, it quickly cools down. Why? Because it loses energy to its surroundings. If I don't want it to lose energy, I have to keep supplying it at a steady rate with new energy, if I don't want its temperature to change. It will always lose energy, but I'm simply replacing the lost energy, and maintaining a constant temperature.

That's the key to maintaining constant temperature. That's the key to keeping warm. You want to balance the energy loss that naturally occurs with energy supply to the system. In things like buildings, we have thermostats that turn the furnace on and off to achieve that balance. In our brains, we have a hypothalamus. It does things like cause us to sweat more. In animals, it raises the fur, ruffles feathers, or whatever, and such way that that affects the energy loss rate or increases metabolic rate, to change the energy input, to keep us at the same temperature.

Even without some active mechanism like that, though, objects naturally seek a certain temperature. An example I'm going to give you to end with here is the example of a solar heated greenhouse. Before I do, let me just give you the general conceptual framework for this idea of energy balance.

Heat flow depends on temperature differences. If you are too hot, the heat flow increases. Well, then there's more cooling. If you are too cold, the heat flow decreases, resulting in a warming. An object will therefore ultimately come to a balance at a constant temperature, if the energy input to that object or system remains constant.

What that temperature is, depends on how rapidly the object can exchange heat with its surroundings, and as an example, here's a greenhouse. This greenhouse is going to be in balance. It's cold outside, but the greenhouse can still be warm inside. Why? Because energy is coming into the greenhouse from the Sun, and energy is going out of the greenhouse at an equal rate. That rate is determined by the material the walls are made of, how well the glass is double-layered to be insulated, and so on.

Exactly what that warmer temperature will be is determined by the properties of the greenhouse and how it has been designed, but the point is that in balance, the rate at which energy is coming into this greenhouse—in this case, sunlight—is exactly balanced by the rate at which energy is going out to the cooler environment. You might say, "Wow, there must be some very complicated, computerized heating system to get that balance just right." Nonsense. There's nothing like that. This happens completely automatically.

What if it were too cool in the greenhouse? What if you opened the door, left it for a while, and you said, "Oh, the door is opened," and closed it? Now, the greenhouse is cool. What does that mean? Because it's cool, it's not at a very big temperature difference from its surroundings, and so, the rate at which energy flows out is relatively small. However, the rate at which energy is coming in from the Sun is still large. That means that there's a net input of energy to the system. Its internal energy is going up. Its temperature is a measure of that internal energy, and therefore, the temperature goes up.

On the other hand, if it's too hot in the greenhouse—let's say that 100 people go in there and do aerobics for a while, and their body heat and so on makes it too hot in there for awhile in the greenhouse—then, the greenhouse is much warmer than its surroundings are, and the rate of heat flow depends on the temperature difference, just as it did for my Styrofoam slab. Therefore, the energy flows out at a greater rate than it comes in from the Sun. That results in a loss of energy, and the greenhouse cools down. Eventually, it finds itself at an equilibrium temperature, which is determined by the rate at which energy is coming in, and by the rate at which it is losing energy through its walls. That's what determines temperature.

The same is true of this light bulb's filament. This light bulb is surrounded by the room, and right now—it is a 200-watt light bulb, and I'm dumping 200 watts of electrical energy into it—it quickly reaches the right temperature so that it is hot enough when compared to its surroundings that it is radiating energy at the rate of 200 watts as well. Therefore, it is at an equilibrium temperature. It is too cool, it's getting more energy coming in than going out, and it warms up. If it is too hot, it's radiating more energy than it has coming in, and it cools down. It achieves a balance.

The Sun is at 6,000 kelvins because the energy input to the surface of the Sun from the nuclear reactions at its core is exactly balanced by the energy radiated away from the Sun into space. Again, there isn't some complicated mechanism that achieves that. It simply happens naturally. The system comes into energy balance, because the energy loss rate, the rate of heat flow, depends on the temperature difference.

That kind of energy balance is the ultimate determinant of Earth's climate. That will be the topic of the next lecture.

Lecture Twenty-Six
Life in the Greenhouse

Scope: Earth's climate is established by the same energy-balance condition that governs a building, a star, or a lightbulb. Earth receives energy from the Sun, mostly in the form of visible light. Our planet returns energy to outer space, mostly in the form of longer wavelength infrared radiation (IR). For a stable climate, energy input and output must be in balance. The details of that balance are determined by a number of factors, especially the composition of the atmosphere as it affects outgoing infrared radiation. Although the atmosphere is largely transparent to the visible radiation from the Sun, it is much more opaque to the outgoing infrared. Thus, the atmosphere acts like an insulating blanket, and as a result, Earth is warmer than it otherwise would be. This so-called *greenhouse effect* is perfectly natural and has been a feature of planet Earth—and our neighbor planets—for eons. But we humans are altering the composition of the atmosphere, leading to an enhanced greenhouse effect and a warming planet.

Outline

I.   Solar energy provides 99.98 percent of the energy input to Earth and its atmosphere. (Nearly all the rest is from within Earth's interior, and a tiny amount of energy comes, via tides, from the orbital motion of the Moon.) For Earth's climate to be stable, the solar energy input must be balanced by energy Earth returns to outer space.

   A.   Because Earth is relatively cool, it loses energy by radiating invisible infrared radiation (recall the lightbulb demonstration from the previous lecture)—as opposed to the visible radiation from the hot Sun.

      1.   Knowing the law that describes energy loss by radiation, it's a simple matter to equate the incoming solar energy with the radiation loss and solve for the temperature. The result is a calculated average temperature for Earth of about 0 °F or -20 °C—certainly in the right ballpark but seeming rather cold for a global average.

    2. In fact, Earth's global average temperature is about 60 °F or 15 °C. Something else must be affecting Earth's energy balance.

B. That something is Earth's atmosphere.

    1. Because the atmosphere is transparent to the visible light from the Sun, most of the incident sunlight reaches Earth's surface. Some is reflected off clouds, and a little is absorbed in the atmosphere.

    2. The atmosphere is substantially opaque to the outgoing infrared radiation from Earth's surface. Specific gases that absorb infrared radiation are responsible for making the atmosphere opaque. Most important among these are water vapor and carbon dioxide.

    3. These infrared-absorbing gases are called *greenhouse gases*, because they function something like the glass cover of a greenhouse, trapping heat within. Actually, the term *greenhouse gas* is somewhat of a misnomer, because the primary function of the glass in a greenhouse is to keep warm air from escaping—thus inhibiting convective heat loss, rather than radiation.

    4. The greenhouse gases act as an insulating blanket. In a house, insulation reduces the amount of energy we need to consume to maintain a desired temperature. But on a planet, the energy input is fixed by the Sun and the planet's distance from it. Earth still must get rid of the same amount of energy—as if you kept the furnace on the same amount of time even after adding insulation to your walls. To get that energy out through the "insulation" of the greenhouse gases, Earth's surface must be warmer than it otherwise would be. That's the *greenhouse effect*.

    5. The *natural greenhouse effect*, caused primarily by water vapor and, to a lesser extent, by carbon dioxide, keeps Earth on average about 60 °F or 33 °C warmer than it otherwise would be. Our planet is a lot more comfortable because of the greenhouse effect!

II. How do we know this greenhouse explanation for Earth's climate is correct? The answer lies in a tale of three planets:

A. We can't do controlled experiments on the Earth (although it's arguable that we're engaged in an uncontrolled experiment as we alter the composition of the atmosphere), but we do have two neighbor planets, each with a very different atmosphere, and we can apply the theory to them. Because we know each planet's distance from the Sun, we can determine how much energy it receives. We know that a stable climate requires that each planet lose through infrared radiation as much energy as it gains from the Sun.

    1. Mars's atmosphere is very diffuse, with less than 1 percent the density of Earth's atmosphere. Consequently, we don't expect much of a greenhouse effect on Mars. Indeed, Mars's average temperature is only a few degrees warmer than a simple energy-balance calculation would suggest.

    2. A simple calculation suggests an average temperature for Venus of around 50 °C or 122 °F—a bit warmer than Earth because Venus is closer to the Sun. In fact, Venus's average surface temperature is nearly 500 °C, or about 900 °F. The huge discrepancy arises because Venus has an atmosphere 100 times denser than Earth's, and it's 95 percent carbon dioxide. Venus has a runaway greenhouse effect that long ago raised the temperature far above what it would otherwise be.

B. These three planets—Venus, Earth, Mars—each in a different way helps confirm the theory of the greenhouse effect.

III. Since the beginning of the industrial era, human activities—mainly the burning of fossil fuels—have increased atmospheric carbon dioxide by about 30 percent.

A. Of the solar energy captured by plants in the process of photosynthesis, a tiny fraction is not cycled through the Earth-atmosphere system but ends up buried in the ground as the fossil fuels coal, oil, and natural gas. The fossil fuels store not only energy but also carbon. It has taken tens of millions of years to accumulate today's reserves of fossil fuels.

B. We're burning fossil fuels at a far greater rate than they're being stored, thus upsetting the balance of the carbon cycle and increasing atmospheric carbon dioxide.

C. Climate data going back hundreds of thousands of years show a clear link between temperature and atmospheric carbon dioxide.

1. Carbon dioxide and temperature have fluctuated together, with much of the time spent in low-temperature, low-CO_2 states called *ice ages*.

2. Every hundred thousand years or so, a briefer (about 10,000-year) warm period occurs. We're in such a warm period now. These fluctuations are believed caused by subtle changes in Earth's orbit, enhanced by feedback effects in the climate system. The average temperature difference between a warm period and an ice age is only about 10 °F.

D. The graph of ancient climate shown on the video gives a maximum of 280 parts per million for Earth's atmospheric CO_2 concentration before the industrial era. Where would today's level be on this graph?

1. Today's atmospheric carbon dioxide is higher than anything the planet has seen in the past half-million years and probably in the past 20 million years.

2. Although this glosses over many subtleties and complexities, that's the main reason for concern that we humans are making substantial changes in Earth's climate.

E. Our best estimates from computer climate models suggest that Earth will warm some 1.5–6 °C (3–9 °F) by the year 2100. That compares with a warming of about 0.6 °C (1 °F) during the 20th century and 6 °C (10 °F)—but in the other direction—between now and the last ice age. Even a few degrees' change is sure to have a significant impact on agriculture, sea level, weather patterns, and ecosystems. Whether or not the physics of Earth's energy balance is important in your life, it will be in your children's and grandchildren's!

Suggested Reading:

W. Thomas Griffith, *The Physics of Everyday Phenomena: A Conceptual Introduction*, chapter 10, pp. 202–203.

Going Deeper:

Stephen H. Schneider, Armin Rosencranz, and John O. Niles, eds., *Climate Change Policy: A Survey*, chapter 1.

Richard Wolfson and Jay M. Pasachoff, *Physics for Scientists and Engineers*, chapter 19, pp. 488–489.

Questions to Consider:

1. Is the greenhouse effect good for the Earth or not? Discuss.
2. A temperature change of a few degrees doesn't seem like much; after all, temperature can vary by tens of degrees in a single day. Why, in the context of past climate change, might a few degrees be significant?

Lecture Twenty-Six—Transcript
Life in the Greenhouse

Lecture Twenty-Six, "Life in the Greenhouse." In the previous lecture, I introduced a number of ideas relating to heat: The flow of heat, temperature scales, how we measure temperature, and so forth. In this lecture, I would like to apply those ideas to planet Earth. The particularly important idea that I would like to apply is the idea of energy balance, which I introduced at the end of the previous lecture. With that idea, I described in particular a greenhouse, heated by the Sun that warmed up and achieved an internal temperature higher than the outdoor temperature. It did so by achieving a balance between the energy coming in by sunlight, and the energy that was lost, in that case, largely by conduction through the walls of the greenhouse to the outside.

That process of energy balance ultimately determines the temperature of any object or system that is subject to an influx of energy, and then loses energy by any of the three energy transfer methods we talked about last time: Conduction, the flow of heat through a material medium; convection, the transfer of energy by the motions of a warm fluid; or radiation, the loss of energy by electromagnetic waves.

Of those three, only radiation can take place in the absence of matter. Therefore, a system like a planet, like the planet Earth, ultimately can lose energy to space only by radiation. Things therefore get a little simpler. We only have to worry about one transfer mechanism when we are talking about the overall energy balance of the planet Earth. We have to be a little bit more careful than that though, because intermediately, between the Earth and the atmosphere, and between different levels of the atmosphere, other mechanisms—particularly convection and the motions of heated air—may play a significant role.

Now, what I'm going to give you in this lecture is ultimately just a brief introduction to the signs of climate, and climate change. It's what could become a whole course in itself, but we're going to just look for a brief interval at some of the key scientific ideas. Now, I'm well aware that climate change is a subject that sometimes seems to be fraught with controversy. I'm going to stick strictly to science here. I'm not going to talk about the consequences of climate change. I'm not going to talk much about the future of the climate change that we expect to happen on Earth, although I will briefly mention some aspects of that.

I do want to say, though, at the outset, in the spirit of this particularly scientific discussion of climate change, is that some of the controversy you may have heard about surrounding climate change is really not much of a controversy when it comes down to the scientific community. Much of climate science is absolutely solidly established, and people of all shades of political persuasion are in agreement about many, many, many of the aspects of climate change. The vast majority of climate scientists are in agreement about the fact that climate change is now occurring, and that it is something that we human beings are at least partly responsible for.

I'm not going to go any further into that, but I wanted to dispel the notion that there still is a very broad controversy about climate change. We have come to know a great deal about Earth's climate, and particularly, in the last few decades, we've come to know a lot, and have firmed up our understanding of climate change. I'm going to all in this lecture only on the scientific basis of climate change, to give you an understanding of what determines Earth's climate, what determines the temperature of our planet, and what might be happening to change that temperature, either naturally, or because of things that we human beings do.

Now, one thing we have to understand about the Earth that greatly simplifies establishing its climate is that unlike some systems, Earth has, almost exclusively, a single source of energy. That source of energy is the Sun. Sunlight provides about 99.98% of all the energy coming to the surface of planet Earth. Most of the rest comes from the interior of the Earth. It comes from the primordial heat that still left over from when the planet formed, 4.5 billion years ago. Some of it also comes from the decay of naturally occurring radioactive elements.

A tiny fraction of the energy coming to the surface of the Earth comes from inside the planet then, and there are a few places—like Yellowstone National Park, or the geysers power plant in California, or the island of Hawaii—where that heat is a significant player in the energy flows in that region. By and large though, that's not a significant player in the energy flow to planet Earth; 99.98% of the energy coming to planet Earth comes from the Sun.

In addition to the energy from the interior, there's a very small amount of energy coming from the tides as they dissipate their energy crashing against the land shores, or simply through fluid friction in the water. That energy ultimately derives itself from the rotation of the Earth, and the revolution of the Moon around the Earth, but that's almost insignificant, with the exception of a few places where we humans have harnessed tidal power to generate electricity.

The easy thing, therefore, to remember about planet Earth is that it gets almost all of its energy from one source, namely, the Sun. Here we have to think of another aspect of something I introduced last time, particularly, radiation. I showed you—I had a 200-watt light bulb here, and a variable autotransformer that I could use to set the amount of electric current through the light bulb. We saw two things. We saw that the hotter the light bulb, the more energy it radiated.

We also saw something else, though. We saw that the hotter I made bulb, the whiter the color of the light was. At first, it glowed a very dull red. Actually, it was glowing mostly in the infrared, but we couldn't see it. It was a little bit in the red. Then, it was orange, then yellow, and then eventually, it glowed kind of white-hot. There was therefore a difference in the wavelength, the color of the radiation emitted by objects, depending on what the temperature was. The Sun emits light that to us appears white, and is peaked in roughly the yellow-green area of the spectrum. That's because the sun's surface temperature is about 6,000 Kelvins, or roughly 6,000 degrees Celsius. We could convert that to Fahrenheit if we wanted to, but I won't bother. The Sun is therefore pretty hot, and that determines the color of the light emits.

The Earth, and the other hand, is a relatively cool object. The surface temperature of the Earth is, in rough, round numbers, 300 Kelvins, about a 20th of the surface temperature of the Sun. That means that the kind of radiation that the warm Earth emits is a very different kind. It's all electromechanical radiation, but it's a very different wavelength of electromagnetic radiation than what the Sun emits. The Earth emits infrared, not even infrared that's very near the color red, but infrared that is much longer in wavelength, and much lower in frequency.

That's important in understanding climate. Sunlight comes from the Sun. It's a very short wavelength radiation. It's mostly in the form of visible light. The Earth, in turn, radiates long wavelength radiation, infrared radiation. Sure, they're electromagnetic waves in both cases. It travels at speeds through vacuums. It consists of crossed electromagnetic fields, as I showed you in a previous lecture in Module Three, but it may react very differently with matter, and that's the key to understanding Earth's climate.

Let's take a look at what Earth's energy balance looks like. Here's Earth, sitting in the middle of space, and again, because it is surrounded by the vacuum of space, the only way to exchange energy with its surroundings to any significant extent is by radiation.

Here comes radiation in from the Sun. Incoming sunlight hits Earth. Basically, it is always hitting just one side of Earth, as the planet rotates. That's why we have night and day. The Earth warms up. If the Earth were not warm, if the Earth was cold at the time we turned on the Sun—this is a simple experiment to think of, but it helps you get a sense of how the Earth gets with temperature—if the Earth were extremely cold—zero Kelvins, absolute zero—sunlight would come in, it would supply energy, and the planet would warm up.

The planet would begin radiating a little bit of electromagnetic radiation at very long wavelengths, because of its very low temperature. It would heat up and emit more electromagnetic radiation. Eventually, it would heat up to the point where it was emitting as much energy as it was getting from the Sun. Then, like greenhouse of the last lecture, it would begin energy balance, and would be at a constant temperature.

We can calculate that temperature very easily, because we know the law that describes how hot objects or warm objects emit electromagnetic radiation. I mentioned last time that they do so in a way that depends on the fourth power of the temperature. I'll show you a formula with that in just a minute, and get a tiny bit quantitative here.

We also know the rate at which sunlight comes in to the planet Earth. I'll tell you that rate. Above the atmosphere—if you put a satellite up there, and we certainly have done this with many satellites—you would measure, on every square meter of Earth's surface, about 1400 watts falling from sunlight on every square meter, not on Earth's surface down here, but above the atmosphere. By the time that radiation gets through the atmosphere, by the way, in a great noonday Sun, a square meter gets about 1,000 watts, about a kilowatt; about as much energy as it takes to run, for instance, a hairdryer. With my 20% efficient solar panel I showed you in several lectures, that would get you out about 200 watts, for example.

However, the entire planet is not in exposed sunlight, and is not in exposed sunlight of such intensity, especially at the poles, or the sunlight is hitting relatively obliquely. If you average overnight and day, and all that, the rate at which sunlight hits an average square meter of Earth's surface, it's just about 240 watts.

That's a good number to remember. Remember when we had Jamie appear, turning the crank on a generator? She was barely able to produce 100 watts. So, every square meter—even averaging for night and day, and the fact that some areas get less sunlight than others—all comes out to an average of about 240 watts on every square meter, enough to run a very bright light bulb, for example. We know that's 240 watts per square meter.

So now we have the Earth in this state of energy balance. In comes the sunlight, and all goes out infrared. I've described the sunlight is yellow, because it's visible. I've drawn the infrared in red, to remind you that it's a different wavelength. Let's do the calculation.

We know the rate at which sunlight is coming in is about 240 watts at every square meter, watts per square meter. We know that the infrared radiation going out is given by a law that I described, sort of, last time. I said that depends on the fourth power of the temperature. Actually, there's some universal constant of nature, and it's given by this symbol, Sigma, sitting here. It says, "Infrared out, Sigma T4." We know the value of that constant Sigma. This is determined by experiment that we can do in the laboratory.

If the Earth is to be in energy balance, that infrared out, Sigma T4—and notice that depends on how hot it is, the hotter it goes, the bigger T is, and therefore, the bigger T4 is, and we get more infrared out—that Sigma T4 has to be equal to 240 watts per square meter. If it is, we are in energy balance. The Earth is losing as much energy as it is gaining from the Sun. If that were not in balance, it would quickly come into balance, because if the temperature were too low, more energy would be coming in. The planet would heat up. If the temperature were too high, more energy would be going out, and the planet would cool down, so that eventually, the planet would reach equilibrium. At what temperature? At the temperature determined by that 240 watts per square meter being equal to Sigma T4.

When students take my climate change course at Middlebury College, one of the first things they do is that calculation, and they get an answer that doesn't seem too bad. The answer comes out to be 255 Kelvins. That's about -18 degrees Celsius, or about zero degrees Fahrenheit. Now, where I live in Vermont, zero degrees Fahrenheit is not unusual. Twenty below zero Fahrenheit is unusual, but if I had to say what the average temperature in northern Vermont was—the year-round average—zero degrees Fahrenheit would sound a little low to me. Most people in the United States, at least, live in warmer areas. Those numbers therefore sound a little low. It sounds as though it's in the right ballpark, but it doesn't seem quite right.

Something else must be going on here. That's the key to understanding the more subtle details of Earth's climate. This picture I have here, and this calculation, give you the really gross overall details, or the gross overall picture, but it isn't quite right. It gives an Earth whose average temperature is a bit on the cool side.

What else is going on? What else is affecting Earth's energy balance? The answer is that what is affecting Earth's energy balance is Earth's atmosphere. The reason the atmosphere can affect the Earth's energy balance has to do with the fact—as shown in the picture that I'm working from—that the incoming sunlight is visible light, shortwave radiation, and the outgoing red is longwave radiation. Those two things in fact react differently with the gases in the atmosphere. In particular, the atmosphere is roughly transparent to visible light. That's why we can see the Sun. That's why we can see the stars. That's why we can see distant galaxies with our telescopes, because Earth's atmosphere is—not entirely, but essentially—transparent to visible light.

On the other hand, it is substantially opaque to outgoing radiation, outgoing infrared. The reason is not because of the dominant gases in the atmosphere, oxygen and nitrogen, but because of some more complicated molecules with three atoms—most of them—particularly water vapor, H_2O, our good old friend water, evaporated in the atmosphere, and carbon dioxide. These are the two gases that have the biggest effect in trapping outgoing infrared radiation and in keeping it from going out. The atmosphere is not entirely opaque, but largely opaque, to outgoing infrared radiation because of those gases.

Those gases are called *greenhouse gases*. The term is a slight misnomer. The idea is that they sort of act like the glass on the greenhouse that keeps the heat from escaping the greenhouse. The reason that is kind of a misnomer is because the main thing the greenhouse glass does is to prevent air from convecting out of the greenhouse, and carrying energy away with it. It has a slight effect in blocking outgoing infrared, but that's not its dominant effect. Its dominant effect is to stop convection, whereas Earth's atmosphere's dominant effect, the greatly significant effect, is to block outgoing infrared radiation. The term "greenhouse gas" is a bit of a misnomer then, and the whole "greenhouse effect" is a bit of a misnomer, but we will stick with it, because that's what we tend to use.

What happens? Here's what we call the "greenhouse effect." Here's the picture of Earth's energy balance, but now, here is Earth with an atmosphere. I've drawn the atmosphere kind of exaggeratedly thick. What happens is that the infrared—the greenhouse gases—act as a kind of blanket, like an insulator. They make it hard for heat to escape. It's a little more subtle than that. They actually absorb the infrared, they heat up, and they re-radiate it, like any hot object does, but they radiate it in both directions.

Some of it goes out to space, some of it comes back to the surface, and as a result, the Earth's surface heats up. It has to be at a hotter temperature than it would have had to be without the atmosphere. In order to keep radiating back to space, that same 240 watts per square meter is what it has to do to get itself in energy balance. As a result, before we humans industrialized, the natural greenhouse effect, a few hundred years ago, resulted in the Earth's average temperature being warmer by about 60 degrees Fahrenheit, about 30 degrees Celsius, than it would have been otherwise. That's why that zero degrees Fahrenheit was kind of low. We are actually 60 degrees warmer. The average temperature of the Earth is about 60 degrees Fahrenheit, or about 15 degrees Celsius.

That's called the *natural greenhouse effect*, and it's a wonderful thing. The Earth would be habitable, but very uncomfortable, if it weren't for the natural greenhouse effect. Again, the natural greenhouse effect is caused primarily by water vapor in the atmosphere and, to a significant but lesser extent, likewise carbon dioxide.

Now before I go on, let me just give you some hint that we know this is correct. This is not just some cockamamie theory that someone has come up with without testing it. This is a theory we know to be correct, and we can test it; unfortunately, we can't test it by trying to modify Earth's atmosphere, and seeing if it's right. We are doing that, but kind of inadvertently. We have a kind of uncontrolled experiment going on, which I will get to more in a minute.

We do have two other planets nearby, for which we can also calculate what the temperature ought to be and compare with what the temperature actually is, and what the constituents of the atmosphere are. We really have what I like to call a "tale of three planets." In this picture, this table, I list our neighbor planets, Venus, Earth, and Mars. Now, they're different distances from the Sun, and so, we would expect them to be at somewhat different temperatures, although that effect isn't all that dramatic. Venus, for example, wants to be at 50 degrees, if you do that simple calculation, sunlight in equals infrared out, 50 degrees Celsius. That's hot, but it's not boiling.

Earth, as we found, ought to be at -18 Celsius, and Mars wants to be at -60, because it's a little further from the Sun. Significant variations, but not dramatic.

What are the actual temperatures? Well, Mars is only a hair warmer than projected. Why? Because Mars' atmosphere, although it's got a lot of carbon dioxide, is only about one one-hundredth the density of Earth's, so that there is essentially no greenhouse effect. Earth is, as we know, 33 degrees Celsius warmer, 66 degrees Fahrenheit. Warmer, because of a fairly significant greenhouse effect. Venus has an atmosphere that is 100 times denser than Earth's, is 95% carbon dioxide, and long ago, it had a runaway greenhouse effect, which brought it up to this cooking temperature of 500 degrees Celsius, about 900 degrees Fahrenheit. Venus has had a runaway greenhouse effect, and its surface is entirely baked because of that.

This is therefore a kind of experiment that nature has provided us, convincing us that this whole theory of the greenhouse effect is in fact correct, and describes planetary atmospheres and planetary climates. Now the question is: What we human beings doing to Earth's climate? In particular, what are we doing to Earth's atmosphere? The main thing that we have been doing for about 200-300 years is burning fuels, fossil fuels that represent trapped sunlight, and that was buried in the Earth long ago. We are burning those fossil fuels, and releasing the products of combustion into the atmosphere.

We have actually altered the atmosphere in significant ways. Here's a graph of carbon dioxide concentration in the atmosphere from roughly 1750 or so to about the year 2000. Actually, it's a little bit beyond 2000. This graph goes up to 2005. What has been going on here?

Well, the atmospheric concentration of carbon dioxide has been increasing, and it has being increasing fairly dramatically, and fairly rapidly in recent decades. It started out at about 270 parts per million. That's a unit that says, okay, if I took one million one-gallon milk jugs, and filled them with here, and separated them into their components, 270 of those milk jugs—not that many of those milk jugs—would contain pure carbon dioxide. That number has risen though by about 30%, until in the middle years of the first decade of the 21st century, it is now around 380 parts per million, and rising.

How do we know that? Well, these data come from a variety of sources, but in recent years, they come from a very careful monitoring station on Mauna Loa in Hawaii, as well as other places. I show in an inset graph here the Mauna Loa data from about 1960. You see it fluctuating up and down, and that's actually due to the fact that in the Northern Hemisphere, when summer comes, plants come out, and they absorb carbon dioxide. The carbon dioxide level goes down, and then, when they lose their leaves, the carbon dioxide level goes back up again. You might say, "Why doesn't that happen in the Southern Hemisphere?" It doesn't, because the Northern Hemisphere has most of the landmass, and we are seeing that asymmetry. I show you that give you a sense that we know what this carbon dioxide has been doing.

We also know that this carbon dioxide is from ancient sources, because we can get it by the amount of radioactive material left it, which is almost none. If it were new carbon dioxide coming out of the atmosphere, the surface of the ocean, or something, it would be more radioactive. More on this in Lecture Thirty-Four. It wouldn't be dangerously radioactive, but it would have a little radiation to detect to date things.

We therefore know that we have changed the carbon dioxide concentration of the atmosphere in recent years substantially. Now, to understand a little more about what effect this might have on climate, we need to look at what climate has been doing over time. Here's a picture of what we think the temperature of the Earth has been like over the last 160,000 years. This comes from data that's been extracted from deep ice. We have drilled into Greenland and Antarctica, we have extracted little bubbles of trapped air from there, and we can measure specifically the carbon dioxide content. By studying the different kinds of oxygen that are in there, we can actually get a rough measure of temperature.

You see a pattern in there, and it's a pattern that goes back at least 500,000 years, because we have ice cores that far back. This particular graph goes from zero, the present, on the right, to 160,000 years ago on the left. However, the same kind of general pattern repeats itself for these 500,000 years, and we have good reason to think that it's roughly for something like 20 million years.

What's the pattern? The pattern is that there are these peaks of relative warmth. The present is one of these periods. They are called *interglacial warm periods*. They last about 10,000 years, and you can see one that occurred in this picture perhaps 120 to 130,000 years ago, and another one is occurring now.

Those are punctuated by much cooler periods, which we call *ice ages*, or *glacial periods*. They are significantly, but not dramatically, much cooler, and one important thing to notice is that in this graph, the difference, the average temperature difference between the warm period we are in now, and an ice age, is on the order of six degrees Celsius. That is something on perhaps the order of 11 degrees Fahrenheit.

It's not a huge difference. Remember that when someone says, "Oh, it's going to warm up five or six degrees. How can that possibly matter?" Well, the difference between now and an ice age, when there were two miles of ice on top of much of North America, down to about Long Island, which is the terrain of the glaciers that they left, after they receded—that difference is a matter of maybe 10 degrees Fahrenheit.

That's what the temperature has been doing, roughly, over the last 160,000 years. It's been repeating this pattern before that. We think this pattern is triggered by subtle changes in the Earth's orbit. It is affected by other planets, by the tilt of the Earth's access, and things like that. Then, natural mechanisms within the climate system cause of a rapid rise in temperature, and then the slow fall.

What happens to the carbon dioxide? Well, the carbon dioxide tracks almost identically, and again, this is carbon dioxide measured from these bubbles in these trapped gases, in these ice cores. You might say, "Well, clearly, there's cause and effect here, but what is the cause, and what is the effect?" It's not that simple, and we don't fully understand the connection completely, but we think what happens is something like this: It warms up a little bit. When it warms up, the ocean can't dissolve as much carbon dioxide, so the carbon dioxide comes out of the ocean.

Various other things like that occur, and that causes it to warm up still more, because of the greenhouse effect. We think a sequence of these feedback effects, which we don't fully understand, causes a very small change in temperature to be amplified into this rapid rise of about 10 degrees Fahrenheit, about six degrees Celsius. It brings us into the interglacial warm period, and then, as the subtle changes occur again in the astronomical parameters, that gradual feedback again gradually takes that down. We don't understand all the details, and we are currently studying a lot of these rapid fluctuations now, which you see in this graph.

Now, I want you to remember a number that I showed you when I showed you the industrial era CO_2. Remember that the concentration started out at about 270 or 280 parts per million. You see that on the right hand edge of the carbon dioxide grass, the blue graph, where I put the scale for parts per million. What is the number today, and where would it be on this graph? Think about that for a minute.

Okay? That number, in case you didn't remember, was about 380 parts per million. Where would that be on this graph? Up there. The top of that arrow is today's carbon dioxide concentration in Earth's atmosphere. That change, which brings us into a régime that our planet has not seen for at least the past 500,000 years—and probably not seen in the past 20 million years—gives us a higher level of carbon dioxide than the planet has experienced in that kind of time. We human beings are known to be the cause of that change, and that's not something about which there is any controversy at all.

I want to emphasize that we have already done to the atmosphere something quite dramatic. We have changed its carbon dioxide level by a huge amount, by at least as much as the difference in the carbon dioxide level between now and when we were in an ice age, but going in the opposite direction. That's a big effect on the atmosphere. That's we human beings affecting the atmosphere globally, and as the developing world develops—particularly China, with its over one billion people in the coming decades—that's only going to go up. China has enormous reserves, and this is its quickest way to industrialize and get energy. We're going to see a lot more carbon dioxide going into the atmosphere.

What's going to happen? What's this doing to our planet's climate? Well, we have good enough temperature records that we can look at the actual temperature, with pretty good confidence, over the last 150 years, since about 1850 or so. Years of graphs of one of several data sets—they look almost identical—in which people have carefully interpreted this data, and have worked on a number of effects that might have given you incorrect readings, and have compensated for them, and so on.

What you see at the start of this period is a fairly constant temperature, with significant rise in the early part of the 20th century, then leveling off or even a period of decline in the middle years of the 20th century, which may ironically being due to the rapid industrialization efforts after World War II, in which a lot of pollution was put into the atmosphere, and actually resulted in a cooling. Then, there was a rapid rise in the final three decades, particularly, of the 20th century. That rise is continuing into the 21st century. This data goes through the year 2003, and it includes the warmest year on record, until that time, the year 1998. The years 2003 and 2002 were also very warm, there, near the top years. There has therefore been a dramatic rise in the Earth's temperature in the past few decades.

The question is: Is this something that might be occurring naturally, or is that related to this dumping of large amounts of carbon dioxide into the atmosphere? Well, we would like to answer that question, but we really can't push this record back further with thermometers, because we don't have temperature records going back further than that.

However, we can reconstruct that temperature. Here's a reconstruction that was done by climatologists, who took into account, in this case, 112 different proxy indicators for temperature—the state of coral reefs, tree rings, cosmic ray data, the ratio of different oxygen isotopes, the kinds of organisms that were found in sediments, all kinds of things—112 different indicators, to try to reconstruct what we think the global temperature. Actually, this is the Northern Hemisphere's temperature, but the global temperature is probably very similar over the last thousand years. Recently, by the way, this particular study has been pushed back almost 2,000 years.

The general trend—the blue there, the reconstructed temperatures. The very faint blues are the error limits. We're 95% confident that the actual temperatures lay within those values, and are reasonably confident that it followed something like that blue and black graph that you see there in the middle. If you had to sum that up, you would say that the first 900 years of the millennium, from the year 1000 to the year 2000, was characterized by a gradual decline in temperature, with fluctuations.

The last 100 years—that is the 20th century—and this doesn't quite go into the 21st, but if it did, that trend would continue—were characterized by a very rapid rise. That's the red data. The red data are the actual data I just showed you from thermometers. The red and blue overlap a little bit, because the reconstruction was carried forward, and indeed agrees with the record from thermometers.

It's therefore likely something dramatic has happened in the last century, something that is dramatic on time scales of at least 1,000 years. Now, that doesn't mean that there might not be natural fluctuations on the level, but that such a natural fluctuation would coincide with the time in which we have done something so dramatic to the atmosphere is most unusual, and that's part of what these climate scientists are stating with considerable confidence, that much of the warming of the last 50 years—this is an internationally agreed-upon statement—much of the warming we've experienced in last 50 years is related to human activities, and in particular, to the burning of fossil fuels.

What do we think is going to happen in the future? Well, the future is harder to predict. We have considerable confidence in our computer climate models, which allow us to—we've actually done models where we've started with conditions 1,000 years ago, and projected to the present, and it works very well. If we put in the human-caused carbon dioxide, it shows that substantial upward rise, but predicting the future is a little bit more iffy, in part because the future depends on what human beings do. If we enacted substantial laws to control the emission of carbon dioxide, for example, that would affect what the temperature will be 100 years from now.

As best we can tell though from computer models, there are a range of temperature rises that we can expect in the current century, before the year 2100 that ranges from roughly 1.5 degrees Celsius to about 6 degrees Celsius—so it's very crudely, perhaps three to nine degrees Fahrenheit.

In that context, if I take this graph of 1000 years of temperature reconstruction and build temperature data, I have to shrink it down considerably to add to it what we think will happen in the next 100 years. I have to change those scales. I'm going to use the middle range production from the intergovernmental panel on climate change, up to the year 2100. That's we can expect might happen, if the middle range of projections—not the extreme ones or the minimum ones, but the middle range ones—turn out to be true. We are going to see a temperature rise in both rapidity and amount that is unprecedented, certainly, over a couple of thousand years, and probably much longer in the history of our planet.

I'm not going to talk about the consequences, but there will be consequences. And even if this isn't something that bothers you, it means that this aspect of physics in your life is going to be something that's very significant in the lives of your children and grandchildren.

Lecture Twenty-Seven
The Tip of the Iceberg

Scope: Things don't just heat up or cool down in response to heat flows. They also expand and contract. More dramatically, they change state—melting and boiling to form liquids and gases or condensing and freezing. A substantial amount of energy is involved in these transformations, and which transitions occur may depend on external conditions, such as atmospheric pressure. One of the most commonplace of substances—water—is unusual in its thermal behavior. Unlike most substances, water expands when it freezes, and consequently, ice floats. This abnormal behavior also affects liquid water near the freezing point, with profound consequences for living things.

Outline

I. Heat flow into or out of an object does more than gradually raise or lower the object's temperature. It also changes an object's dimensions and, more dramatically, may cause a change of state.

II. Most substances expand when heated. This is the result of the increasing energy of individual molecules, which interact more violently when heated and, therefore, maintain a slightly larger spacing.

    A. For typical solids, thermal expansion is a fairly small effect, amounting to only about one 1/1000 of a percent change in length for each degree change in temperature.

        1. Engineered structures need to account for thermal expansion or disaster can result. Bridges, parking garages, and similar structures are equipped with expansion joints to allow for thermal expansion without cracking the structure.

        2. Precision instruments, such as telescopes and other optical systems, are often built from special materials designed to minimize thermal expansion. Otherwise, even small changes in temperature could compromise performance.

        3. Bonding together two materials with different expansion rates gives a structure that bends as its temperature changes. This effect is widely used in such applications as thermostats and automatic greenhouse vents.

B. Liquids generally show more thermal expansion than solids, and gases, even more.

    1. Thermal expansion in a column of liquid makes for a simple, accurate thermometer.

    2. Thermal expansion in closed systems can be dangerous. If you have a water-based home heating system, it better be equipped with an expansion tank to accommodate the extra volume as the water heats up. Your water heater has safety relief valves that discharge water if pressure builds up from thermal expansion, thereby preventing a possible explosive failure of the water heater. Your car's cooling and fuel systems also have devices to handle thermal expansion of coolant and fuel.

    3. One projected consequence of global warming (discussed in the previous lecture) is a rise in sea level. Roughly half of that rise will be from thermal expansion of the ocean waters (the rest is from melting ice).

C. Although at most temperatures water expands when heated, very near its freezing point this most common of substances is unusual. Between 0 and 4 °C (32 and about 39 °F), water's volume actually decreases as it warms. Thus, water is at its densest at 4 °C (about 39 °F).

    1. This unusual behavior means that water at 4 °C sinks to the bottom of lakes, and in many deep lakes, the bottom temperature remains at this level year round. In the summer, temperature rises toward the surface, but in the winter, it drops. Either way, the lake is stable, with less dense water on top of denser water.

    2. However, twice a year, in spring and fall, the lake surface warms or cools through 4 °C. At that point, denser water overlies less dense water, and the lake "overturns," in the process churning up nutrients from the bottom and generally revitalizing the lake.

III. A more dramatic thermal effect involves changes of state of a substance—from solid to liquid, liquid to gas, or even solid to gas—that occur abruptly at particular temperatures.

A. A substantial amount of energy must be added or removed to effect a change of state (also called a *phase change*). For example, it takes nearly as much energy to melt a chunk of ice as it does to raise the resulting water from the freezing to the boiling point. And it takes more than five times as much energy to boil the water away, turning it all into a gas.

B. While a substance is changing state, its temperature doesn't change. Once ice is brought to its melting point (0 °C or 32 °F) and begins to melt, the ice/water mixture stays at that temperature until the ice is all melted. Only then can the temperature begin to rise again. On cooling, water reaching the freezing point (again 0 °C or 32 °F) stays at that temperature until it's all frozen.

C. Although it need not be any warmer than ice, liquid water contains more energy—the energy that was added to melt the ice. Similarly, water vapor contains a lot more energy than liquid water, even if they're both at the boiling point (100 °C or 212 °F). When water freezes to ice or water vapor condenses to liquid water, that energy is released.

  1. This energy "stored" in the "higher" state is sometimes called *latent heat*, because it can be released by changing a substance to the "lower" state.

  2. Even the slow process of evaporation that occurs below the boiling point requires energy. (The boiling point is special because here, the pressure of the evaporated vapor is equal to atmospheric pressure.) Evaporation is, therefore, a cooling process—a fact that underlies the operation of your household refrigerator (more on this in the next lecture).

  3. Latent heat plays a major role in weather and climate. As solar energy evaporates ocean water, the air becomes both warmer and moister. This warm, moist air rises, taking with it the energy that went into transforming liquid water into vapor. Higher in the atmosphere, it may re-condense to form clouds—in the process, releasing its energy. The energy released from latent heat is what powers tropical hurricanes and explains why they quickly lose strength when they move over land.

D. The temperature at which state changes occur depends on pressure. That's why water boils at lower temperatures at high altitudes, and it's why pressure cookers and nuclear power plants can heat water to higher than 212 °F without it boiling.

  1. A *phase diagram* summarizes the relation between solid, liquid, and gaseous phases at different temperatures and pressures. At a fixed pressure, such as Earth's atmosphere provides, many substances show all three phases. But at other pressures or for other substances at normal atmospheric pressure, there may be only two phases. That's the case for carbon dioxide, which *sublimes*, changing directly from solid "dry ice" to gas.

2. The phase diagram shows two other interesting features that, for most substances, fall outside the realm of "everyday physics." Above the *critical point*, liquid and gas become indistinguishable, with the substance making a gradual transition from dense liquid to diffuse gas as the temperature increases (for water, this occurs at 374 °C and a pressure more than 200 times atmospheric). Fluids near the critical point exhibit some remarkable properties, many of which are best seen in the apparent weightlessness of an orbiting spacecraft. And there's a special point, the *triple point*, where solid, liquid, and gas can all coexist. That point occurs at a unique temperature, and thus, it provides a rock-solid way to calibrate temperature scales.

E. Once again, water is unusual. Unlike most other substances, its solid phase is less dense than the liquid; therefore, ice floats. This has profound implications for aquatic organisms; if ice were denser than water, lakes would freeze from the bottom up and aquatic life would be difficult. Instead, ice forms an insulating layer on the surface, allowing life to continue beneath. Our planet would be a very different place if water behaved like most substances.

1. The reason for water's unusual properties lies in the crystal structure of ice. Individual water molecules (H_2O) bond hydrogen to oxygen to make a very open structure with a lot of empty space—hence the low density. Incidentally, this is why snowflakes are six-sided.

2. This unusual structure means that ice can be made to melt under pressure. This helps you pack loose snow into a snowball.

3. A residue of this bonding effect keeps the molecules in very cold liquid water farther apart than they otherwise would be—hence water's unusual property of expanding when cooled at temperatures below 4 °C.

4. Because ice is only about 10 percent less dense than water, it floats low in the water (recall Lecture Ten)—which is why all we see is the "tip of the iceberg."

Suggested Reading:

W. Thomas Griffith, *The Physics of Everyday Phenomena: A Conceptual Introduction*, chapter 10.

Paul Hewitt, *Conceptual Physics*, chapter 17.

Going Deeper:

Richard Wolfson and Jay M. Pasachoff, *Physics for Scientists and Engineers*, chapter 20.

Questions to Consider:

1. When you emerge from swimming, you feel quite cool until you get dry—even on a hot day. Why?

2. You add some ice to a glass of water and wait a while. When you come back, there's still ice in the water. Is the water warmer than the ice? Discuss.

Lecture Twenty-Seven—Transcript
The Tip of the Iceberg

Welcome to Lecture Twenty-Seven, "The Tip of the Iceberg." In this lecture I will answer the question I raised in my lecture on flight about the floating of things, and why icebergs float with most of their bulk below the water, as well as many other questions involving the thermal behavior of materials. What happens to materials as we add heat to, or take heat away from them?

The most obvious thing that happens—and we've already discussed this—is that when you add energy or not—one way to add energy is by heat, which is a flow of energy that occurs with the temperature difference—the object heats up or cools down, depending on whether there's a net gain or loss of energy. That's how we established the temperature in a greenhouse, and then, in the previous lecture, that's how we determined Earth's climate.

Other things happen though when temperature changes, or when heat is added, to a substance. It may, for example, change shape. It may change size, or more dramatically, it may change state. Most substances, for example, when you heat them up—meaning when you raise their temperature—expand. The reason they expand is because ultimately, that thing that we call, loosely, "heat," and which I argued is more quickly called "internal energy," involves the rapid vibrations in collisions of particles making up the substance, and if the temperature is higher, that vibration is more vigorous. That causes the particles to tend to stay just a little bit further apart from each other. That's why things tend to expand.

For materials that are solid—hunks of steel, concrete, solid things, wood, whatever—that thermal expansion, the expansion due to temperature changes, is pretty insubstantial. It's typically about 1000th of a percent for every degree Celsius temperature change, and is not a big effect. Nevertheless, it's enough of an effect that sometimes, it's important. In some engineered structures—for example, bridges—if you go across a bridge, you'll see metal plates that interlace like your fingers do here, like my fingers are doing, at the start of the bridge. That's so that as the bridge expands, there's a place for it to go, because the bridge that is a mile or so long, even though it expands much less than a percent, is nevertheless a significant distance over a mile.

Railroad tracks, especially continuously welded railroad tracks have to be carefully distressed so that the thermal expansion doesn't cause them to warp. In fact, there's a very famous picture that appears in many introductory physics texts—here it is on the screen—the famous event in which a very hot day caused thermal expansion of these railroad tracks. These railroad tracks really had nowhere to go, so they buckled sideways, and you see the train in the background, derailed.

Another related disaster—not quite thermal expansion, but thermal properties and materials—was the 1986 shuttle disaster with Challenger, which exploded because the O-rings, the rubber rings that were used to hold together the different sections of the solid fuel rocket, behaved differently in different temperatures. They became stiff, particularly at the low temperatures the morning of that launch. That's what caused the Challenger disaster.

Sometimes, this thermal expansion, in precision instruments like telescopes, for example, or optical benches for careful scientific optical measurements, that tiny, tiny change can be a problem, and if that happens, we often build such devices are very special alloys that are designed to have a very minimal amount of thermal expansion compared to more ordinary materials.

Sometimes, we would like to exploit that property, to make useful things happen. Many thermostats—for example, old-fashioned thermostats, devices that are used to open vents on greenhouses automatically, many other devices that respond in a simple mechanical way to temperature—we achieve a more significant response by binding together to different materials. I have here a thin strip of metal, about a half an inch wide, and about a half a millimeter in thickness. It really consists of two different metals are bonded together. Those two different metals expand at different rates when they are heated, and when the one that expands more expands, it forces the whole structured event, and that bending gives us a significant movement that can be used as either a measure of the temperature, or maybe a close on an electrical switch, or something like that.

I'm going to do a quick experiment with this device. I'm going to start by warming it over a candle, so I will light my candle. Notice that there is now a slight curve to what is your right. I will try to hold it in that same orientation.

A candle, by the way, although it produces high temperature, doesn't put out a lot of energy. Again, the difference between temperature and heat, or temperature and energy; temperature is a measure of the average molecular energy, but the number of molecules affected by this candle is pretty insignificant. Gradually though, the strip is warming up.

You can see that now, it has been substantially—or a little bit—in the other direction. Now, I'm going to take it and plunge it into a thermos of very cold material, and it comes up at the other way. By the way, that was liquid nitrogen, at 77 Kelvin, very cold. The vapors you see streaming off are not the liquid nitrogen, but water from the air condensing on contact with this very cold rod, and making, basically, a small cloud.

There you have the use of materials with a so called *bimetallic strip*, with two different kinds of materials that expand at different rates, and can bend this way if it's heated, this way if it's cool. Again, a device like that could be used, for example, to close an electrical switch when a certain temperature is reached, and that's how old-fashioned thermostats work, although many of them have been supplanted by thermostats that use electronic thermostatic sensors of the type I described in the module on semiconductors.

There's an example of a bimetallic strip. Here's a very simple example of where a bimetallic strip might be used. This is a simple, inexpensive thermometer. You see the needle that indicates temperature, and what the needle is ultimately connected to is simply a coiled up spring. That spring is made of a bimetallic material, and as the material changes as temperature, the spring either coils tighter, or uncoils, depending on which way it goes, in the same way that the strip of bimetallic material swung one way or the other, as I heated or cooled it.

We can therefore exploit this property of materials' thermal expansion to cause motions that tell us something about temperature, or open and close circuits, turn furnaces on and off, or whatever.

That's the thermal expansion of solids. Liquids and gases tend to expand more, liquids a little bit more—several powers of 10 more—and gases substantially more. I just want to give you some examples of that. By the way, here I have a couple of glasses of ice water. Keep these in mind. They will occasionally appear on your screen, when the cameras are pointing at them, and notice that one of them is completely full to the brim. It's got lots of ice floating in it. We're still pretty near the beginning of the lecture, and just remember that fact. I'll get back to that at the end.

The other one is only partly full, and I've got a thermometer stuck in it. That thermometer happens to be reading zero degrees. More on that later, but the fact is that that thermometer—it's an alcohol-based thermometer; you can see the red alcohol in the bulb at the bottom—that thermometer relies on the thermal expansion of that fairly large bulb of material, and as it expands up the narrow hollow tube of the thermometer, a little bit of expansion into volume results in a fairly large change in the height of that column of alcohol. That's how a simple thermometer like this works.

Thermal expansion in gases though is more dramatic. I'm going to demonstrate that with this balloon, which is full of a gas at room temperature. I've got it in a glass pan, and the glass pan is resting on a couple of pieces of Styrofoam, just to insulate it from the table. I'm going to pour liquid nitrogen on it.

Now, liquid nitrogen is not intrinsically dangerous as a material, because it's just air, basically, without the oxygen, which is really the more dangerous part of air. With the oxygen removed, Nitrogen comprises 80% of the air, so when this stuff boils away, it's just part of the air. However, it's extremely cold, and could freeze my tissues instantly, so I'm going to take some precautions. I'm going to put on big gloves, and I'm going to put on this mask to protect my face if any of it should splatter. Those are good precautions when one is handling a material this cold. By the way, your dermatologist uses liquid nitrogen to remove warts by freezing them off.

All I'm going to do now is put on the facemask and the gloves. I've got the liquid nitrogen in a thermos bottle, and I'm going to pour it over the balloon. The balloon is going to—now remember, the balloon is at room temperature that's about 300 Kelvin, and liquid nitrogen is at 77 Kelvin. That's about one quarter of the temperature of the balloon, in absolute terms. So, the balloon's volume is really going to change substantially, and this will be a good example of thermal expansion of a gas.

Well, the balloon has certainly lost most of its volume. It's sitting there in the liquid nitrogen. By the way, the rubber has also lost most of its flexibility, an example of what happened in the Challenger disaster. You can see the vapors flowing over and off of the table. They are flowing downward because they're cold compared to the surrounding air. I can feel cold air on my hands here. Again, those vapors are not the liquid nitrogen. Those vapors are the water vapor in the air condensing on contact with the very cold air surrounding here, and they are then condensing, making these clouds, and the flow of that cold, dense air is made visible by the presence of the water vapor forming clouds.

Now, the liquid nitrogen is exposed to room temperature, and so, heat has been flowing into it, because it was at 77 Kelvin. It was surrounded by a room and air at 300 Kelvin. The liquid nitrogen has all boiled away, and the balloon is beginning to regain some of its flexibility, and it's also beginning to warm up. By the end of the lecture, it will probably have warmed back up to room temperature, and will look perfectly normal again.

There's thermal expansion then, and a gas that is quite dramatic, especially if you change the absolute temperature by a substantial amount, as we did there by dropping it down to 77 Kelvin. By the way, you probably have systems in your house—I certainly hope you do—that protect against thermal expansion, because it can be dangerous, especially in gases and fluids in things like boilers. Early steam engines and steam boilers tended to blow up with disastrous losses of life. Your house probably has in it a heating system—at least if you're in a cooler climate—that involves heating water, or some other substance. You don't want that to expand and cause the plumbing to burst.

Here's an example of the kind of thermal protection that you probably have in your house. You don't have exactly this because unfortunately, you probably don't solar heat your hot water. I do, and these are the solar collectors on the roof of my house. There are six of them. The middle one looks different because it's a skylight. There's a lot of material—it's actually not water, but a kind of antifreeze, a nontoxic antifreeze, because in the winter, when it's 20 below, I wouldn't want water to freeze and expand, and crack the pipes.

This material circulates, and it's heated. As it gets up near the boiling point of water on a hot, sunny day, it expands, and needs to have somewhere to go. You'll find that in your basement, if you have a hot water heating system, or a solar hot water system as I do, you'll find an expansion tank like this, whose job is to have a place for the extra volume of the heated liquid, in this case, this fluid that circulates and collects heat from the solar collectors, and transfers it to a tank of water. That's what that tank is doing, and you'll have one like that if you have a hot water heating system for the same reason.

You probably also see in your basement structures that look like this. This is a relief valve. This particular one is on the side of my solar hot water tank. Its purpose is to blow if the pressure gets above a certain value, or if the temperature gets above a certain value, in this case, 30 pounds per square inch, because if the water in that tank got so hot that it expanded, it would push against the valve, and finally, for the expansion tank, in particular, there would be no place for it to go, and the valve would go. I have occasionally gone away on very hot weekends in late August and very early September, sunny weekends, and my solar collectors are optimized for that time. I have come back to find the basement flooded, and not much hot water in the tank, because this valve has let go. There are therefore plenty of devices that give us protection against thermal expansion.

By the way, I talked about global warming in the previous lecture. One possible consequence of global warming is a rise in sea level. Most people think that will occur because the ice will melt. Well, some ice will melt, and polar ice, land ice that melts will contribute to the rise in sea level. Ice that floats in the water will contribute nothing to the rise in sea level, and that's what this glass jar is demonstrating. As the ice melts here, you do not see water flowing over the sides. That ice, because it's already floating ice, and its density will change when it turns to water, will not contribute anything to sea level rise. About half of sea level rise due to global warming though, will simply be the thermal expansion of the oceans. They're acting like giant thermometers.

Now, as I said at the beginning, most materials expand when heated, and water is no exception, except that there is a temperature range from about zero to four degrees Celsius—32 degrees, the melting point, to 39 degrees Fahrenheit—at which water behaves in a very unusual way. Water actually contracts upon heating at those temperatures. It's a very strange phenomenon. It makes water a very unique substance.

I will give you a sense a little bit later about why that happens, but it has an unusual consequence for life, aquatic life particularly, because it what it means is this: It means that if water is very deep—it means that water at four degrees Celsius, or 39 degrees Fahrenheit is the densest that water can be. If you therefore have water at that temperature, it sinks to the bottom of any body of water, and in any lake that's very deep, the water temperature at the bottom is usually about four degrees Celsius, 39 degrees Fahrenheit, year-round.

In the summer, it's much hotter at the surface, so let's take a look at what a lake looks like. Here's a lake in the summer. At the top it might be 24 degrees Celsius, 75 degrees Fahrenheit. The bottom is four degrees Celsius, 39 degrees Fahrenheit. The top is warmer and less dense, so I have less dense water sitting on top of more dense water. It's perfectly stable, and it stays that way. If you've gone into a lake, you'll find that it gets colder as you go down. Usually it's not a gradual temperature change, by the way, but there's often a sudden layer in which the temperature changes significantly. By the way, my balloon it's almost back to normal.

What happens in the winter? Well, in the winter, there's usually a thin layer of ice on a lake. Just below the ice, the temperature is typically as cold as water can get and be liquid water. That's zero degrees Celsius or 32 degrees Fahrenheit. At the bottom of a lake, it's still four degrees Celsius. The densest water is still at the bottom. The zero-degree Celsius water, the water just at the freezing point is less dense because of this unusual behavior of water, and everybody's happy.

Twice a year though, in the spring and fall, as the water heats up in the winter, the temperature at the surface will rise through four degrees, and a lake will have essentially the same temperature throughout. In the fall, as the surface temperature falls from its warm summer temperature down through four degrees, you'll come to another time when the temperature in the lake is pretty much uniform throughout. At that point, the lake is unstable, and the water from the bottom can mix.

You get this phenomenon called *lake overturning* or *turnover*. That's a very important phenomenon in lakes, because it results in bringing up nutrients and things that are near the bottom of a lake, and returning them into the lake. It kind of keeps the lake revitalized. So, this phenomenon of lake turnover, which you know about if you live near lakes, is something that occurs because of this thermal effect, this fact that water has the unusual property that it actually can contract when heated for the first four degrees Celsius of its watery, liquid existence.

Now, more dramatic thermal effect involves the changes in state of the substance—turning a liquid into a vapor, turning a solid into a liquid, turning a liquid into a solid, etc. These changes require substantial amounts of energy. It takes a lot of energy. That's why if you want to melt ice, it takes a long time to melt ice. It takes almost as long to melt an ice cube—almost as much energy to melt an ice cube—as it does to bring it up to the boiling point. It takes a huge amount of energy from then on to boil it. In fact, the energy that it takes to melt an ice cube is about as much energy as it takes to raise the temperature of that water after melting to about 80 degrees Celsius, almost to the boiling point.

Here's something that people often don't understand: While a substance is changing state, it remains at a fixed temperature. Water, just melted, is at exactly the same temperature as the ice it came from. It has more energy, and the energy went into breaking the bonds that hold the ice together into a solid structure. However, water, just after melting, is not any warmer than ice is. This glass of ice water is at exactly the same zero degrees Celsius, 32 degrees Fahrenheit. I don't even have to look at the thermometer to tell that, because I know that a mixture of ice and water— and atmospheric pressure that I have to add—is, in fact, automatically at zero degrees Celsius.

As long as there is both ice and water together in there, the temperature of that mixture will be zero degrees Celsius, will be exactly at the freezing point. Only after it all becomes water can it begin to rise. Until then, any energy coming into that from outside is not going into making it warmer, but is going into changing its state to melting ice. If I put this in a freezer, it would stay at zero degrees Celsius, even if the freezer were much colder, until all of the water had turned solid. At that point, the temperature could start to go down.

There is therefore energy involved in changing state of a substance. Water has more energy than ice does. Water vapor has more energy than liquid water does. Sometimes that energy is called *latent heat*. It is called "latent heat" because it is sort of latent in there. You can get that energy back if you change the state back again.

Processes that require things to evaporate, resulting in a net change of state, require energy. For example, if I do this simple experiment—which you are welcome to do as well—I'll put a little water on my hand, so that my hand is wet, and blow on it. I've caused the water to evaporate. I have caused the fastest moving water molecules in that water to leave my hand, and that process takes energy away from the water. The remaining water is cooler, and that's why my hand feels cold. That's why you can dip your hand into water, or lick your finger, hold up to the breeze, come and see which way the breeze is coming from, because the breeze evaporates the water, and your hand cools down. That's what you're really feeling. It takes energy to change the state of something.

Latent heat plays a major role in weather and climate. It's one of the energy transfer mechanisms from the Earth up into the atmosphere, because if the Sun warms the waters of the ocean, water evaporates, goes into the atmosphere, and the warmed air rises. It carries with it not only the water vapor, but it also carries the latent heat, the energy that it took to change the water from liquid to gas. Further up, as it gets cooler, that water can re-condense, form clouds, a visible manifestation of condensation, just like clouds that water formed here from around the liquid nitrogen when it was cold, but in addition, that energy gets released, and some of the most powerful storms, particularly hurricanes—here's the aerial photograph of a hurricane, from an orbiting satellite; you can see the curvature of the Earth in the distance, and you can see the big spiral structure of the hurricane—the energy of that storm, the colossal winds of that storm, the big swirling motion, are fed by latent heat associated with the evaporation of water from the warm oceans. That's why hurricanes are tropical storms that form in tropical waters. It's also why hurricanes, as soon as they drift over land, tend to die down.

Now, we think of water has changing state at 32 degrees Fahrenheit, melting from ice to liquid, and we think of it boiling at 212 degrees Fahrenheit, but that happens to be true because of the particular value of the pressure of the Earth's atmosphere. Water boils at very different temperatures, in, say, a pressure cooker—more on that in the next lecture—or in the pressure chamber of a pressurized hot water reactor nuclear power plant. I'm going to give you a very quick and simple demonstration of water boiling at a different temperature than its normal 212 boiling point. This is certainly what happens on top of a mountain. You may be aware—if you live somewhere like Denver—that the temperature at which water boils is lower, because the atmospheric pressure is lower.

Here I have a little jar, from which I can take some of the air out. I can't boil water at room temperature with this simple apparatus, but what I am going to do is boil some coffee, which is already hot, but not boiling. I'm a tea drinker, so I don't know whether this coffee is any good, but let's pour a little coffee into here, not boiling, but hot. I have a little vial of coffee. I'm going to put it on the bottom of this jar. I'm going to put the jar over it, and now, with this very simple crude vacuum pump—it's not a big electrical one, so it can't make all that much of a vacuum—I'm going to pump out some of the air in that chamber.

If we watch closely, after a couple of pumps with this thing—there's a one-way valve; I pull the air out, and then I exhaust the air out in another direction—there it goes, boiling vigorously. Still, the temperature it was at, perhaps 180 degrees Fahrenheit; wasn't the boiling point of water. The boiling point therefore depends on the pressure that's applied. In general, those temperatures at which something changes depend on pressure.

We can summarize that relationship in what is called a *phase diagram*. This is getting a little more quantitative, but I want you to see this, because I want you to understand that water, which you think of as ice, liquid water, or vapor, is only one of many possibilities for a substance. This is a graph. On the bottom axis, we have temperature increasing, on the right; on the vertical axis, we have pressure increasing, to the left. Here's a particular curve that I will talk about in a minute, and here's another curve. These two curves constitute a phase diagram for a particular substance. By the way, it isn't water. That red curve would be sloping a different way if it were water, because of that unique property of water that we've talked about.

To the left of the red curve, the material is a solid, regardless of what pressure you're at, as long as you are above that very low pressure on the blue curve. To the right, you are in a region where the material is definitely liquid. Beyond that, you are in a region where the material is a gas. A substance like water, at normal atmospheric pressure—that might be normal atmospheric pressure, if this were the graph for water, which it isn't, quite, but it could be—there's a slight change. As you change the temperature, keeping something at atmospheric pressure, it goes from solid, to liquid, to gas; but there are other ways to make that transition. I just made the transition from liquid to gas in that coffee by moving vertically on this diagram, and crossing the liquid/gas boundary that way, by lowering the pressure, while the temperature stayed the same.

The normal sequence that you think of solid, liquid, gas, with increasing temperature, is only an artifact of the particular pressure that we have on Earth, and the particular substances we are used to.

Carbon dioxide, for example—if I had some dry ice here, the dry ice would never melt at normal atmospheric pressure. It would simply turn directly from solid to gas, a process called *subliming*. If this were carbon dioxide I were talking about, then that line would be normal atmospheric pressure, and at normal atmospheric pressure, carbon dioxide is not a liquid. In your fire extinguisher, it is a liquid, but there, it's under very high pressure. As soon as it comes out, it turns into a gas, and sometimes a little bit of dry ice forms around the edges as it cools rapidly upon expanding.

Other things can happen. For example, at very high pressure, we have solid, and then, that curve of liquid to gas kind of ends abruptly. It ends at a point called the *critical point*, and above the critical point, there is, in fact, no real distinction between liquid and gas. The material goes gradually from a very dense liquid to a lesser dense liquid, and just turns gradually, without that abrupt transition, into first a dense gas, and then, more and more diffuse gas.

The critical point is unusual. We don't think about it, because the critical point occurs for normal substances in temperatures and pressure that we don't tend to get to. For water, for example, it's at 374 degrees Celsius, and about 200 times atmospheric pressure. That's not a place we go frequently. Fluids that are very near their critical point exhibit some really remarkable properties that are under active study today, and some of them, in fact, are best seen in orbiting spacecraft.

I should point out one of a place on this diagram that is special. It's right there. It's a point at which those two curves meet. At that point, liquid, solid, and gas can all coexist in equilibrium. Right here, in these two containers, I have liquid and solid coexisting. This water is at some point along the liquid to solid curve. At any point along that curve, you can have both liquid and solid in any arbitrary amounts, and they can happily coexist together. That's anywhere along that curve. If I changed the atmospheric pressure, the temperature in here will change, the temperature at which that transition occurs.

However, there is one unique place—the triple point—where all three phases can exist together. If you can get a material to its triple point, you know absolutely what its temperature is. For water, the triple point happens to be 273.16 Kelvin, and that, in fact, is used to define the scale of temperature. We have absolute zero at the bottom, and if you put water at its triple point, that, by definition, is 273.16 Kelvin. You can't be wrong, because there's only one unique temperature, regardless of pressure—well, it's also a unique pressure. You have to have that pressure and that temperature, and you will be at the triple point, but the temperature will be known. For water, it's 273.16 Kelvin.

Once again, water is unusual. Water is unusual because its solid phase is less dense than its liquid phase. That means that ice floats. If water were like most substances, when winter came, ice would form, and it would sink. However, that doesn't happen, and that has a profound effect, obviously, for aquatic life, because instead, the ice forms an insulating layer, which allows life to go on, perhaps, at a reduced pace in the cool water below the ice. If ice sank though, lakes would freeze from the bottom up, instead of from over the top, and we would have a very different aquatic life. In fact, it's arguable that some forms of life may not have been able to exist at all.

The reason ice is less dense is somewhat unusual. It has to do with the crystal structure of ice. Here's a picture of the crystal structure of ice, the molecular structure of ice. You see in this picture, pictures of the water molecule, the H_2O molecule, with the two "H"s on either end, and the bigger "O" in the middle, forming an angle of 105 degrees. I mentioned in the electricity module that ice and other substances are sometimes held together by unique bonds that you see here as dashed lines between the hydrogens, the positive hydrogens, and a neighboring oxygen, in a neighboring water molecule. That forms this very hollow kind of structure, and it is the hollowness of that structure, with a lot of empty space, that causes ice to have a lower density than water does.

By the way, it is the residue of that structure—the residue of that bonding effect—that still exists even after ice is melted for the first few degrees, that causes water to have that unusual behavior of becoming still more dense as it warms up. Is because of that residual attractiveness; but beyond about four degrees Celsius, water begins to behave perfectly normally. You see this hexagonal structure to the crystal structure of ice; that's what results directly in the hexagonal structure of snowflakes. No two snowflakes are the same, but they all share this kind of hexagonality, which you can see here, this six-sidedness.

Long ago, I showed you a salt crystal. I poured a few pieces of salt out and then showed you a cubicle salt crystal, and showed you the structure of the NaCL, the salt crystal underneath. It's the same with water except that water's hexagonal structure leads directly to these hexagonal pictures for ice.

By the way, the fact that ice has this unusual property means that you can make ice melt by putting it under pressure. Normally, if you put a substance under pressure, you would solidify it. For ice, though, you can melt it. That's one of the things that allow you to pack a snowball together, and keep it together, because you actually melt the ice with the pressure to put it together.

Now, ice is only about 10% less dense than water, and that means that it floats very well in water. That's the answer to the question that is implicit in the title of this lecture, "The Tip of the Iceberg." Because ice is only 10% less dense than water, it floats very low. By the way, that value depends on where the ice came from. Ice from the Antarctica has more rocks entrained in it, so it has a different density than ice from the Arctic.

Finally, let me end by taking a look at this glass of water with ice in it. Some of the ice has melted in the course of this lecture, but none of the water has spilled over. That's because floating ice, as it melts, becomes more dense water in exactly the right proportion. That's why it floats at the level it does. It doesn't change the overall volume at all, and that's why global warming, and the melting of Arctic sea ice, although it may have major effects on climate, will not change sea level.

Lecture Twenty-Eight
Physics in the Kitchen

Scope: We cook food to enhance its flavor and texture and to kill harmful bacteria. Before cooking, we store many foods under refrigeration. Refrigeration inhibits bacterial growth and slows natural chemical changes. Cooking is the opposite, intentionally changing the physical and chemical structure of the food. Cooking is essentially a heat-transfer process, in which the inflowing energy alters food properties. Common cooking processes differ in how the energy is transferred, what parts of the food are most affected, and how rapidly and how much the food's properties change. Understanding the subtle differences between the several distinct ways of applying heat to food is one of the marks of a gourmet chef!

Outline

I.   Most foods are best stored at temperatures just a few degrees above the freezing point of water. Low temperature slows the enzymatic chemical reactions that alter the food's properties. It also inhibits the growth of bacteria—both those that spoil the food and pathogens, which, although they may have no obvious effect on flavor, can result in serious disease when ingested.

   A. Even the rough difference of 30 °F between room temperature and the interior of a refrigerator has a dramatic affect on the time food can be kept without spoiling. Fresh milk, for example, keeps about two weeks at refrigerator temperatures; it would spoil in a few hours at room temperature.
      1. The remarkable effect of a few degrees' temperature difference lies in the molecular energy that is the basis of heat-related phenomena. At any temperature, there's a wide range of energies, with the average being an indicator of the temperature.
      2. The relatively few molecules with the most energy, however, are most responsible for driving chemical and biochemical reactions. Raising the temperature slightly doesn't change the average energy very much, but it greatly increases the number

of high-energy molecules. Hence, reaction rates increase dramatically—including food spoilage.

B. Domestic refrigerators work on a principle discussed in the previous lecture—namely, that it takes energy to change the state of a substance from liquid to gas.

    1. The refrigerator contains a *working fluid* whose liquid-gas transition occurs at a temperature somewhat below the lowest temperature desired in the refrigerator. Early refrigerators used ammonia, a material that was hazardous if it leaked. In the 1930s, a synthetic chemical called Freon replaced ammonia. In the 1970s, scientists discovered that Freon and related chemicals destroyed the protective ozone layer high in Earth's atmosphere, and by the 1990s, the use of Freon was phased out in favor of newer, less harmful synthetic materials.

    2. A motor-driven pump compresses the working fluid, at this point in gaseous form, raising its pressure and temperature. The fluid then passes through tubing exposed to the environment outside the refrigerator. This cools the fluid, which condenses into a liquid.

    3. The high-pressure liquid passes through an expansion valve, basically just a constriction in the piping, greatly reducing its pressure.

    4. The low-pressure fluid passes through tubing exposed to the interior of the refrigerator. As it does so, it evaporates. The energy required to change the fluid from liquid to gas comes from the refrigerator's contents, which, therefore, cool.

    5. The fluid is once again condensed and the process is repeated. During the condensing process, the energy that was extracted from the refrigerator contents is transferred to the refrigerator's surroundings. Thus, the refrigerator "pumps" heat from its interior to its surroundings. Unfortunately, it consumes energy in the process—much more about this in the next lecture.

II. Conventional cooking systems are much simpler than refrigerators— for profound reasons that I'll cover in the next lecture. Electric ranges convert electrical energy into heat by passing current through resistive wires. Gas ranges use the heat released in combustion of natural gas or propane. Either way, the purpose of cooking is to alter food characteristics to enhance flavor and texture and to kill harmful bacteria.

A. Cooking involves a number of complex chemical and physical changes. Proteins in meat and other animal products coagulate when heated, making the food firmer and, if carried to extremes, tough. Reactions between sugars and amino acids generate hundreds of different flavor-enhancing substances. Heating, especially in the presence of water, breaks down complex sugars into simpler molecules and, at high enough temperatures, results in caramelization. An important goal of cooking is to exert some control over these processes. Different cooking methods use combinations of the three basic heat-transfer mechanisms—conduction, convection, and radiation.

B. In baking, boiling, simmering, deep-fat frying, and similar methods, food is immersed in a hot medium—air, water, or oil—and energy flows from the medium into the food until it reaches the desired temperature. In all these cases, energy is deposited on the outside of the food, then makes its way to the interior by conduction.

1. In baking, electric elements or gas flames heat the air in the oven. Convective air circulation transfers energy throughout the oven, heating the air and oven walls. Convection also carries energy to the food. Because air has a low heat capacity, the air near the food drops in temperature as it gives energy to the food. Air's low heat capacity is the main reason that baking is a rather slow process. In *convection ovens*, a fan forces a more rapid air circulation, keeping the air near the food warmer and, thus, reducing cooking times. Rapid convection also results in a more even temperature distribution in the convection oven.

2. In boiling and simmering, conduction carries heat from an electric or gas burner through a pan and into water. Convection in the water then carries heat to the food. Because of water's great heat capacity, boiling is more rapid than baking. Boiling in an uncovered or loosely covered pot ensures that the water temperature is the boiling point, or 212 ˚F (100 ˚C). In a pressure cooker, the higher pressure results in a higher boiling point—typically 250 ˚F, with double

atmospheric pressure; therefore, food cooks faster (recall the phase diagram of the previous lecture).

3. In steaming, food is immersed in water vapor over boiling water. Energy is transferred to the cooler food as water vapor condenses on the food and gives up its latent heat. Again, the temperature is held at water's liquid/gas transition point of 212 °F.

4. In a double boiler, excellent for gently cooking egg custards, a second pan holding the food is suspended above boiling water. Water vapor condenses on the bottom of the pan, giving up its latent heat and maintaining a uniform 212 °F. Because it involves no violent convective motion, this is a much gentler method of cooking.

5. Deep-fat frying is similar to boiling, but because fat can be heated well above the boiling point of water, cooking times are typically much shorter.

C. In grilling and broiling, heat is transferred directly to the food.

1. In panbroiling, stir-frying, and griddle cooking, the hot pan itself is the heat source, and heat transfer is by conduction into the food.

2. In broiling and grilling, heat transfer is by radiation from hot coals or an oven's heating element.

III. Microwaving is an entirely different cooking method.

A. Strictly speaking, microwaving itself does not involve heat. Pause and consider why not. Remember the precise definition of heat: energy being transferred as a result of a temperature difference. In microwave cooking, there's no temperature difference; the source of the microwaves (the magnetron; recall Lecture Eighteen) is not particularly hot compared with the food. Microwaving is, therefore, a *nonthermal* method of energy transfer.

B. Microwaves cook because the oscillating electric field of the microwaves acts on molecules that, although electrically neutral, have uneven distributions of electric charge. The water molecule has a particularly pronounced charge distribution, with the oxygen atom more negative and the two hydrogen atoms more positive. Thus, water and water-containing foods are particularly efficient absorbers of microwave energy. Many other substances, such as glass, most plastics, and paper, are not—which is why you can boil water in a paper cup in a microwave oven.

1. The microwaves in an oven are electromagnetic waves of a specific frequency and, therefore, wavelength. The frequency is about 2.4 billion cycles per second (2.4 gigahertz), and the corresponding wavelength is about 12 centimeters or 5 inches. The walls of the oven, behind the typical plastic interior, are metal, so the microwaves reflect and fill the entire oven. The reflected waves interfere, producing regions of constructive (high microwave intensity) and destructive interference (low intensity). If the wavelength were very short (as it is for light), these would be so close as to be unnoticeable. But because the wavelength is significant compared with the size of common food items and the oven itself, regions of constructive and destructive interference are typically a few inches apart.

2. This interference pattern could lead to uneven cooking of food. That's why food in a microwave oven rides on a rotating platform—the idea being to ensure that no part of the food remains long in a "cold spot." In some ovens, the food is stationary, but a rotating metal reflector above the roof of the oven "stirs" the incoming microwaves to keep the interference pattern changing.

3. Microwaves are kept safely in the food by a metal screen embedded in the glass window. As long as the holes in the screen are small compared with the wavelength, very little microwave energy escapes.

4. Unlike other cooking methods that deliver energy to the food surface, microwaves actually penetrate the food. Typical penetration depth is about half an inch. Microwaves heat the food to about this depth, and conduction from the heated outer layer transfers heat further in.

5. Many foods benefit from cooking at low power, which in a microwave oven is accomplished by turning the microwaves alternately on and off. This gives thermal conduction time to distribute the energy more evenly throughout the food.

6. Microwaving is not particularly effective on ice, because the water molecules in ice are locked into the structure of the ice crystal and cannot easily respond to the microwave electric field. That's why thawing foods in the microwave is done at

low power, with the microwave energy being absorbed by other molecules, then transferred by thermal conduction to the ice.

7. Microwaving is very efficient, in that virtually all the microwave energy generated in the oven ends up in the food; there's no heat escaping to the environment. But a typical oven converts only about half of the incoming electrical energy to microwaves, so its overall efficiency is less.

C. Microwave cooking entails some unique dangers.

1. The electric fields in microwaves cause electric currents to flow in metals, resulting in heating that may actually melt the metal.

2. At sharp corners, electric charge can accumulate on metals, resulting in huge electric fields and sparking that can damage utensils or the oven itself or even start a fire. Even something as small as the metal staple on a tea bag can cause problems.

3. An empty microwave oven is hazardous to itself; with nothing to absorb the microwaves, the energy ultimately returns to the magnetron, where it may cause damage.

4. Because microwaving heats water from the outside in, boiling water in a microwave does not result in vigorous convection. It's possible for the water to become *superheated*, exceeding the boiling point but not boiling. The slightest disturbance— like removing a mug of water from the oven—can then trigger dangerously explosive boiling.

IV. The perfect soft-boiled egg: a simple kitchen task?

A. Egg-white proteins begin to coagulate at 63 °C (about 145 °F), while yolk proteins coagulate at about 15 °F higher. A perfect soft-boiled egg must be heated so that its white exceeds 145 °F but not by so much that the yolk gets a lot hotter.

B. Heat transfer into the egg depends on its size, its initial temperature, and the water temperature. Although this isn't a mathematical course, it's amusing to present a scientific formula giving the time for the yolk to reach a given temperature. Good chefs instinctively know this without doing the math!

Suggested Reading:

Louis A. Bloomfield, *How Things Work: The Physics of Everyday Life*, pp. 299–302.

Robert L. Wolke, *What Einstein Told His Cook: Kitchen Science Explained.*

Going Deeper:

Peter Barham, *The Science of Cooking.*

Questions to Consider:

1. Why should you leave the oven door ajar when broiling?
2. It takes exactly twice as long to cook two portions of food in a microwave oven as opposed to one but not significantly longer in a conventional oven. Why the difference?

Lecture Twenty-Eight—Transcript
Physics in the Kitchen

Lecture Twenty-Eight, "Physics in the Kitchen." I guess this lecture is the closest you're going to get at The Teaching Company to a cooking show. Before I start, let me say in deference to my colleagues and other scientists that there is certainly a lot of chemistry and biochemistry in the kitchen. It isn't just physics. However, I'm going to emphasize the physical science aspect of it, being a physicist.

Ironically, in the kitchen, we tend to want to do one of two things to foods. We either want to cool them, or we want to heat them. Rarely do we want to leave them at room temperature, and so, I want to explore in this lecture our means of first cooling foods, and why we do that, and then, looking at our various means of cooking them, ending with one of the most modern, the microwave oven.

Most foods keep better in the refrigerator—not all, but most. Why is that? Well, there are two things going on in foods. There are enzymes that are involved in, ultimately, spoiling food, and causing the food to over-ripen, and, if you will, causing lettuce to get wilted and spoiled, and things like that happening to the food. Those enzymatic chemical processes are slowed down by foods being in the refrigerator. More importantly, there are bacteria, some of which cause food spoilage and make the food look this dreadful, and some of which may not leave an obvious trace, but that may be pathogenic. They may grow and multiply in the food, and cause human disease. We therefore like to keep most of our food cool.

You might think that a few degrees—and it really is a few degrees, a few tens of degrees—you might think that wouldn't make such a big difference, but it does. You might especially think it doesn't make a big difference after hearing me go on and on at the start of this module about heat and thermodynamics, about the fact that if we really think naturally about temperature, we ought to be using the absolute temperature scale, and on the absolute temperature scale, at which room temperature is about 300 Kelvins, or 450 Fahrenheit, the difference between the temperature in the refrigerator in the temperature outside in the room seems pretty negligible. Why is it, then, that we cool foods? What is the physics of food spoilage?

Let's take a the graph I have on the screen here—another slight veering into quantitative, but not too quantitative—and what I want to graph here is that for a typical substance, at a typical temperature, I want to ask the question, "How fast are the various molecules moving around in that substance?"

Now, I've emphasized that the internal energy that we loosely call "heat" but is really better called "internal energy," and which is ultimately a measure of the temperature—the internal energy per molecule is what temperature is really measuring—that energy is not the same for all molecules. There is a spread in energies. Some average energy—and that's what the temperature is measuring. But in a system at some temperature there's a wide spread of molecular energies, and therefore, molecular speeds.

Thus, this particular shows, on its horizontal axis at the bottom, the speed of molecules. These are actual speeds that you might typically have for molecules that these temperatures; and on the vertical axis, I'm going to plot the number of molecules that have a particular speed, or maybe are in a particular small range of speeds.

Here's what the graph would look like—this blue curve—for a substance at the temperature of the refrigerator. There's a peak somewhere. That's the most likely temperature, the most likely speed you are to find. It's around 400 meters per second in this particular case. It's pretty fast, but that is typical of thermal speeds in substance. The average speed, by the way, is not quite equal to that most probable speed, because this graph isn't symmetric. So, if your average it, you get a slightly different speed, probably a little bit to the right. Nonetheless, the point is that there's a peak. There's a typical speed, and normal speed, or a speed that you are most likely to get the molecules at, and they're a little bit different, but they're near that peak.

However, there's a possibility that some molecules will have very slow speeds, even down towards zero—not much possibility below zero. There aren't many at zero, or close to zero, but as you rise toward that typical speed, there are more and more molecules. Then, as you go above that typical speed, the number of molecules falls off again, and there's a kind of tail, a very high-energy tail of molecules with very high speeds.

There aren't very many of them, but they're important, and the reason they're important is that chemical reactions occur more rapidly at higher temperatures because of the higher energy of molecules. Well, there's a single, fairly low temperature. In this case, it's refrigerator temperature characterizing these molecules, but the ones at the tail of that distribution are more likely to participate in chemical reactions, are more likely to cause chemical reactions to occur, because they have higher energy, so if they bang into some other molecule, it's more likely that there will be some interesting action. A molecule breaks apart, atoms form in a new way, chemistry occurs.

Although molecules have this typical distribution and speeds, therefore, the highest speed molecules, although they are in the minority, and there aren't very many of them, play a major role in mitigating the chemical reactions that occur with the substance.

That's the curve of number of molecules versus speed for refrigerator temperature. Now, have argued that room temperature isn't much different, and it isn't. If I plotted the same curve—and these are actual plots from the actual equation that you would use to describe this—for room temperature, it looks almost the same. It's almost indistinguishable. The peak has moved a little bit to the right, because at room temperature, the average speed of molecules, the probable speed of the molecules, is just a little bit higher.

The curve has also dropped down. The head of the peak has a little bit lower. That's because the entire curve has broadened out a bit. The whole breadth over which speeds you're likely to find molecules at the slightly higher temperature is broader, so the molecules are spread over a broader range of speeds, and that's why you see the peak having less height. That's another, secondary effect. Not only does the plot shift to the right, but also the distribution spreads out a little bit. It therefore doesn't look significantly different, consistent with my argument that room temperature and refrigerator temperature aren't much different.

Let's now look at this high-energy tail, though. Let's blow that up just a little bit—in fact, a lot. If I blow up those two curves at the high-energy tail, I see a remarkable difference between the blue curve, the refrigerator temperature, and the red curve, room temperature. There are far more molecules—oh, in this case, depending on the exact value you pick along there—but in the typical range on the left end, there are four or five times as many molecules in the red curve at that speed as there are in the blue curve. Even though there was only a slight shift in temperature, and it changed the overall curve slightly, it changed the high-energy tail dramatically.

Therefore, a change of only a few degrees, warming of only a few degrees changes the number of molecules in that high-energy tail substantially, and it's those molecules that are participating in chemical reactions, including spoilage, including the reactions that go on inside bacteria and allow them to multiply, and that's why a difference between room temperature and the temperature in the refrigerator is substantial. If you put fresh milk in the refrigerator, it will keep there for two weeks. If you put the same fresh milk out on a counter, on a warm day, it will keep only a few hours. The difference is much more dramatic than the difference between, say, the 275 Kelvins and 300 Kelvins between the refrigerator and room temperature countertop would suggest. It's because of this greatly increased number of molecules at the high-energy tail of this distribution.

We therefore need to cool our food. How do we do that? We do it with refrigerators. In the old days, a refrigerator was an insulated box, and in the winter, people went out and cut ice off lakes and stored it in buildings, icehouses, insulated with sawdust. You went and got ice, kept it in the refrigerator, and that kept your food cold.

However, refrigerators invented in the early 20th century—electric- or gas-operated refrigerators—made keeping foods cool a much easier thing to do. What a refrigerator does is to work on the principle, fundamentally, of state changes. I argued in the last lecture that when you change the state of the substance, particularly when you change a liquid to a gas, it takes energy to do that. I showed you one example of refrigeration. I poured some water on my hand, and blew on it. That evaporation cooled my hand, and that is the principle, ultimately, of the refrigerator.

The refrigerator contains a fluid, called a *working fluid*, whose job is to change state and absorb energy from the contents of the refrigerator. In the 1930s, the working fluid was ammonia, a rather hazardous substance which, when it leaked out, caused all kinds of toxicity problems. In later times, a synthetic chemical called "Freon" was invented. Freon is a seemingly inert chemical that is non-toxic, seems to be harmless to anything, and Freon was used for many, many years in refrigerators, until it was discovered in the 1970s that Freon causes chemical reactions in the upper atmosphere that destroy ozone and let additional ultraviolet down into the surface of the Earth.

The Montréal Protocol of 1987 is an international agreement that basically phased out Freon and related compounds in favor of materials that are less harmful to the environment. Although they still act as greenhouse gases, they don't have this ozone-destroying effect. By the way, that discovery in the 1970s led to a Nobel Prize in chemistry, which was the first Nobel Prize awarded for what was basically environmental research.

How does a refrigerator work? Let's look inside a refrigerator. Here's a picture of what's going on inside a refrigerator, and I'm basically showing you some metal tubing—aluminum tubing, some kind of metal tubing—and a couple of devices, a compressor, and an expansion valve. Let's just see how this thing works.

The working fluid is sometimes a gas. I've designated that by red. It's sometimes a liquid, and I've designated that by blue. What happens? There's a compressor. The compressor is a pump that pressurizes the working fluid, and pumps it around in this loop. When the compressor gets hold of the working fluid, it compresses it, and it becomes a high-pressure gas. Then, that fluid moves through some coils—sometimes in the bottom of the refrigerator, which is actually a rather poor place for it in terms of energy efficiency. A better place is on top, and you may have seen ancient refrigerators that actually have a little round thing on the top. That was a better place to put these coils.

Sometimes they're in the back, and sometimes they're in the bottom, but the purpose of those coils is to take that high-pressure gas, and let it get rid of some of its energy, in the process condensing to a liquid. As it goes through those coils, it gets rid of some of its energy, it rejects that energy to the outside environment, the room the refrigerator is in. Translation: If you touch the coils of the refrigerator, either underneath or in back, they're warm. They're giving heat out to the room that the refrigerator is in. A refrigerator, ironically, gives off heat—well, not so ironically. We'll see why in a minute.

Now, we have the working fluid in liquid form, but it is still at high-pressure. It has rejected the heat to the surrounding environment. Then, it went to the expansion valve as a high-pressure liquid, going through this valve, which is basically just a tiny hole, a tiny little orifice, the gas is forced to go through, and as it goes through, it expands and cools and becomes a low-pressure liquid.

Now, we have low-pressure liquid, and low-pressure liquid is then circulated through a series of coils that are insulated from the outside environment, but they're in contact with the innards of the refrigerator. Heat flows—again, that's the right term, because heat is energy flowing, because of a temperature difference—heat flows into these cool coils, and it evaporates the liquid. It takes energy to evaporate the liquid, and that energy is ultimately extracted from the materials inside the refrigerator. It goes into the working fluid. It is pumped around. The heat has been extracted from the contents of the refrigerator, and now the working fluid is a gas, because it's evaporated at the expense of the energy of the things inside the refrigerator, it goes back to the compressor, and the cycle starts over again.

The net effect of the refrigerator is simply to pump heat from inside the refrigerator, from the contents, to the outside. That heat that you feel coming out from the refrigerator—some of it at least; not all of it, unfortunately—is coming from the contents of the refrigerator. You put in some milk that you accidentally got warm in the car bringing it home, and refrigerator brings it back down to refrigerator temperature. The energy it extracts from the milk ultimately comes out through this rather complex cycle that I just described. It comes out through those coils, and it is rejected, basically, to the environment as waste heat.

By the way, this is sort of how heat pumps work, which heat houses, except that they cool the outside environment, and pump heat into the house. That seems all well and good. However, there's something that I've left out of this picture. The refrigerator doesn't work unless you plug it in, and that's a profound statement. This process of heat transferring from the cool stuff in the refrigerator to the warm outside in the environment does not happen spontaneously. It requires a source of energy. You therefore can't get this process to happen for nothing, even though energetically, it might seem possible. It isn't, and that's the entire subject of the next lecture called, "Like a Work of Shakespeare." We will go there, and understand why this is such a profound statement, in the next lecture.

Refrigerators are therefore complicated things, and they're tied into complicated, and actually philosophically pretty profound, statements about the way the Universe works. That's tied up in the fact that you have to plug your refrigerator in.

Cooking appliances, on the other hand, are dramatically more simple than refrigerators. The difference between cooking appliances and refrigerators is night and day. Refrigerators are, in some sense, quite profound. Cooking appliances are quite simple. Conventional cooking systems—not microwaves; we will get to them later—use either the sort of brute fact that if you pass electric current through a substance that has electrical resistance, the substance heats up, and we know why it heats up. From the module on electricity, we learned that the process of electrical conduction involves electrons moving through a material, banging into the ions of the material, giving up their energy, and the whole material ends up in vibrational motion, which is, essentially, heat. Resistance basically converts nice, ordered, electrical energy into random heat, for reasons that I will show you in the next lecture. It's a lousy way to heat things, in some ways. That's how electric stoves work. That's how little electric immersion heaters work. That's how electric broilers work. That's how electric grills work. If you have a gas range, that's just combustion of the gas. Again, I'll show you in the next lecture why that's actually the more efficient way to go, although you do tend to fill your house with the products of gas combustion, which are water and carbon dioxide, mostly, but maybe some other things—natural gas propane—it doesn't matter; you are basically producing heat from the chemical reactions involved in the combustion of the gas.

Either way, the purpose of the cooking is to alter the food's characteristics, one, in ways that make it taste better and be more palatable, and two, to kill harmful bacteria. Both of those goals are goals of the heating process that cooking entails.

In cooking, there are actually a number of complex chemical and physical changes. If you have meat, eggs, dairy, other animal products, they have proteins that tend to coagulate when they're heated, so they make the food's differ. A well-done steak is stiffer than a rare steak, for example. If you push that process too far, the meat or other animal product becomes tough as those proteins coagulate.

There are sugars and amino acids in foods, and reactions among those sugars and amino acids occur in different ways and at different temperatures, and result in hundreds of different substances that have been identified by, literally, kitchen scientists as flavor enhancing substances. So, you want to develop those substances when you cook. If you heat in the presence of water, particularly, you tend to break down complex sugars into simpler substances. In particular, you may bring on the sugars, and you may bring on the process of caramelization, where sugars turn that beautiful golden brown before they burn up completely.

An important goal of cooking—and a goal of the chef, whether he or she understands all this physics and chemistry—is to make these processes work to get the desired result. A gourmet chef, somehow, has an instinctive feel, without necessarily understanding all of this physics and chemistry, for how to bring on the right flavors, as well as what temperatures to apply for what length of time and so on. That's what makes a gourmet cook.

Somehow, though, you have to get energy into the food, and you do that by one of three basic heat conduction mechanisms we've talked about already. In many mechanisms—baking, simmering, boiling, deep fat frying—what you are ultimately doing is surrounding the food by a fluid, which is being transferred from some other source, like the stove burner, or the surrounding oven or whatever. Then, the energy is transferred to the food, ultimately, by conduction. The energy, in this case, reaches the surface of the food, and then it travels by conduction into the interior of the food. In fact, all conventional cooking methods deliver energy only to the surface of the food—I will say more about microwaves later; they are a little bit different—but conventional cooking methods deliver the energy to the surface of the food, and from there, it's conducted inward. It travels inward by conduction, unless the food is a liquid, and then, there may also be convection in it.

In baking, what you do is heat the oven. You have electric heating elements or flames. The entire oven walls become warm, or hot, and the air inside the oven achieves the temperature that you've got the oven set at. That air is in contact with your roast, or your food, or whatever you've got in there, and that warmed air transfers energy into the food. The problem with that is that air has a relatively low heat capacity, unlike, say, water, so that the air that is in contact with the food kind of cools down rapidly. It takes a while for more energy to conduct through the air, and for the convection currents that are set up in the oven to bring warmer air there. Baking therefore tends to be a long, slow process.

If you have a convection oven—there's that word "convection"—the convection oven uses, instead of natural convective motions, forced convection. It has a fan. The fan basically pushes the warm air in the oven into contact with the food, and doesn't let it linger in contact with the food long enough to cool down. That is why convection ovens, although they are still using the process of baking, do so much faster.

On the other hand, if you have boiling or simmering going on, you have the food immersed, or partly immersed, at least, in water. The water is typically near or at the boiling point, and since water has a much higher heat capacity, it's much harder for the cool food to cool down the water. Furthermore, water is a much better conductor of heat, and so, the processes of boiling and simmering take place much more rapidly. Convection carries the energy from, say, the bottom of the pan to the food. There, again, it conducts into the interior of the food.

If you are boiling in an uncovered container, you're guaranteed that the temperature of the water will not exceed 212 degrees Fahrenheit, or 100 degrees Celsius. That's because, again, as I showed in the previous lecture, once the phase change starts, the material has to stay at the same temperature until it is all gone. Woe to you who boil all the water away, because if you have water there, it's at 212, as long as it's at atmospheric pressure.

Now, I showed you in the last lecture how I could boil coffee at a temperature less than 212, if I was at less than atmospheric pressure. Similarly, if I put more pressure on, the water will rise to a higher temperature as a liquid before it boils. That's what goes on if you like to use a pressure cooker. It's a great thing for cooking beets. They're really tough, and you've got to get a lot of energy into them to soften them up, as with potatoes, sometimes.

A pressure cooker is just a big pressure vessel, with a lid that is tight fitting. You put the lid on, and there's a little regulator thing that sits on top, regulating the pressure at typically twice atmospheric pressure. That temperature, water boils at about 250 degrees Fahrenheit, and so, food cooks that much faster, because the water is much higher. That's how a pressure cooker works. By the way, the pressure cooker is also equipped with an expansion safety valve. It's that little rubber thing with the metal piece in it, and if it gets too hot, that thing melts. The steam escapes, rather than blowing up the pressure cooker, which happened with earlier pressure cookers.

If you like to steam food, the food is immersed in water vapor, over boiling water. The water vapor has a lower heat capacity. It acts a little bit more like baking, in the sense that it's a slower, more gentle process, but nevertheless, the food is surrounded by the steam at 212 degrees, and again, it's always at that transition point. The steam eventually rises, hits the top of the pan, condenses, flows back down, and you have a continuous cycle going in the steam. The water vapor is ultimately transferring the heat to whatever is being cooked.

One of my favorite cooking tools, and one that is getting harder to find, because it's not used as much, is the double boiler. Here's a pan in which you put perhaps an inch of water. Here's a pan in which you put something like an egg custard, or one of my favorite desserts, tapioca pudding. The egg custard, or whatever it is, is cooking, due to the heat that is being transferred from the water boiling below, making water vapor.

The water vapor condenses on the bottom of the pan. It carries energy in the form of that latent heat, gives up the energy gently at the bottom of the pan. The bottom of the pan, again, is kept at that transition temperature, 212 degrees, but there is no violent agitation of the custard as there would be with boiling. It's a very nice, gentle, even method of cooking. You can't possibly burn the stuff, and you don't subject it to that kind of vigorous motion that might tough things up, so I really like a double boiler, and that is a method similar to which—sort of the way in which a hurricane works. I argued that a hurricane works by latent heat of water evaporated from the ocean, released in the air to drive the hurricane. Similarly, in the double boiler, that latent heat released from the water that's boiling because of being on a hot stove burner, is then released in contact with the bottom of the upper pan, and that's what causes the food to cook. That's a double boiler.

If you deep fat fry—I wouldn't recommend that method for health reasons, but if you do—the oil boils at a higher temperature than water, so that you can heat the surrounding liquid to a higher temperature, and deep fat frying is a very quick method of cooking.

If you grill or boil, the heat is transferred directly to the food. Stir frying, or pan-frying—what is sometimes called *pan-broiling*, where you throw a slab of meat on a hot pan, or something—is direct conduction from the pan's surface to the food. If you use a charcoal grill, or you broil in the oven, the dominant method of heat transfer in those cases is radiation from the red-hot surface onto the surface of the food. That's why it is particularly easy to burn food with those methods.

All of those conventional methods of cooking, then, ultimately deposit energy at the surface of the food, and then, it works its way in by the process of conduction. Microwaving is entirely different. Strictly speaking, microwaving is not a process involving heat. Think about that for a minute. Remember what heat is. Heat is energy that's flowing as a result of temperature difference. Well, in microwave cooking, there's no such thing as a temperature difference. There is that magnetron tube that is spinning electrons around, and generating its microwaves at 2.45 GHz' frequency—I described that more in Lecture Eighteen—but the magnetron is not hot compared to the food. The energy is not flowing because of a temperature difference. It's flowing because we generated microwaves in a magnetron, and beamed them into the food.

Microwaving is a non-thermal way of cooking. That sounds like a contradiction, but it isn't. Microwaves "heat" the food without any heat. There's energy transfer, but it's not strictly heat, because it isn't flowing because of temperature difference. As I pointed out in previous lectures, the microwaves cook, ultimately, because the water molecule, in particular, has a big separation of electric charge, positive and negative, and the oscillating field of the microwave—which you saw in that little movie I showed you, of how an electromagnetic wave consists of oscillating electric and magnetic fields—that oscillating electric field, going back and forth 4.5 billion times per second, grabs water molecules, juggles them all up, and makes that random motion that we call, loosely, "heat,", but that we should call thermal energy. The microwave therefore deposits energy in the food, and that energy becomes thermal, what we call "thermal energy" or "internal energy," in the food.

Let me just show you an example. Here's some water. It is at room temperature here. It's actually a little below room temperature, about 55 degrees Fahrenheit. I put it in the microwave oven, turn on the microwave oven, and give it a couple of minutes. That water will reach the boiling point quickly. No heat is being transferred. Energy is being pumped into the water, but not heat.

Now, the microwaves in an oven, as I mentioned, have a specific frequency. It's about 2.45 billion cycles per second, about 2.4 GHz. That corresponds to a wavelength of about 5 inches, 12 centimeters. That results—as I argued way in Module One, when I was talking about wave interference—in the possibility that as the waves bounce around the microwave oven—even though you see a plastic interior to the oven, right behind that is metal. The microwaves are bouncing around. There was a little water spilled, and a bubble of it just boiled, which I guess you didn't hear that noise, but I did, plopping noise, that it was caused by.

As the microwaves bounce around, then, there are regions where they interfere constructively, and regions where they interfere destructively. This picture is a kind of symbolic look at what the microwave intensity in the oven might look like. There are hot spot and cold spots. Again, it's the microwave oven—if the microwaves had a wavelength like that of visible light, very, very tiny, this would not matter, because you wouldn't be able to tell the difference between a hot spot and cold spot, they would be so close together; but because the microwave wavelength is about this long— that's because the frequency that's generally for water molecules is associated with that wavelength—we have this problem of forming hot spots and cold spots in the microwave oven.

What do we do? Well, inside this particular oven, the food is going around. I can open it, by the way. That's a sign that it's not a thermal heating mechanism. There's no heat to come out, except for the fact that the water is now a little hotter. There's a rotating plate in the bottom here. That rotating plate, which you can see here, carries the food around, and it goes between the regions where the cold spots and hot spots are.

In some microwave ovens, as I have said before, I believe, there is—let's just go again, a little bit more—a rotating plate that spreads the microwaves out instead of doing this. That's microwave cooking, and the microwaves are safely kept in the food. You might say, "Well, how come the microwaves aren't coming out and filling the whole room with microwaves?" Well, let's stop it again. I think we have got enough that we can see that this water is hot. That water is very hot. If I stick the thermometer in, it goes shooting right up. I don't know if you can see that, but there it goes. It is rising rapidly, because that water is heated, but by a non-thermal means.

If you look in the front of the microwave, there's a glass door so that I can see into it, but if you look carefully, there's a little screen in that glass door, with little holes in it. Those holes, although they are big enough for us to see through, are small enough that they block microwaves. Remember in a very early lecture, I talked about what it meant for a surface to be rough or smooth. I meant rough or smooth compared to the wavelengths of the waves involved.

Well, these holes are enormous compared to the wavelength of visible light. The light goes right through them. However, they are small compared to the wavelength of microwaves, which are about that big, so basically, they block microwave radiation from coming out. That's why you can look safely into the microwave oven, although it's not a good idea to stand in front of it all the time. However, that basically blocks the microwaves from coming out of the oven.

By the way, you want to cook some foods at a lower power level in the microwave, and when you set a lower power level, you'll hear your microwave turning on and off rapidly, to lower the effective power level. That gives the heat time to spread throughout the food, and it cooks more gently. Unlike other cooking methods, the microwaves actually penetrate about half an inch into typical food. There is therefore heating of the outer half-inch, roughly, and from there, the food conducts inward, as it would in a normal cooking process.

Microwaving, by the way, is not particularly effective on ice, for the reason I showed in the last lecture. Water molecules in ice are locked into a crystal structure. They aren't free to undergo the jiggling, and so, when you thaw food and microwave, you're actually grabbing other molecules, warming them up, and then energy is transferred to the ice. That is part of the reason why you don't blast full power if you are thawing, or if you have a thaw setting. It uses much lower power.

Microwaving is very efficient. All the microwaves' energy ends up in the food. That has two implications. One is efficiency. Although it takes a lot more electricity to make, say, 900 watts of microwaves, than it does 900 watts, the oven itself isn't all that efficient, but the cooking process is. It also means that if you double the amount of food in the oven, it will take exactly twice as long to cook it. If you have a package of food that says, "Cook for six minutes," and you put two of them in, you put them in for 12 minutes, because all the energy goes in, unlike the other processes, in which most of the energy is wasted.

Microwave cooking entails some dangers. The electric fields cause currents to flow. They can heat up metal. At sharp corners of metal, you accumulate electric charge, and sparks can fly. Here's this Teaching Company mug. It's got some gold leaf on it. I wouldn't want to put this in a microwave oven, because I would get lightning strikes, literally, on that gold leaf.

Because microwaving heats water from the outside instead of boiling it from below, it's possible that water can heat above the boiling point and not boil. Then, if you pull that cup out, jostle it a little bit, or drop a tea bag in it, it may boil explosively, and splatter hot water on you. There are therefore some dangers with microwave ovens.

Well, I want to end on a totally different topic, just on a little bit of humor. How do you cook an egg? Well, just to show you that this is a scientific proposition, here's Professor Williams' famous egg formula, and I give you at the bottom a web site where you can go look up Professor Williams' treatises on eggs. He's from England.

Here's the cooking time for an egg. "D" is the diameter in millimeters. Notice the cooking scale is the square of the egg diameter. "T" is the hot water temperature. "T naught" is the initial egg temperature, and "T yolk" is the desired yolk temperature. What do you want? Soft and runny? 150. Soft gel? 165. Hard-boiled, 170. You are going to get it turning green at 175, and dry and crumbly at 195. There really is a lot of physics in your kitchen.

Lecture Twenty-Nine
Like a Work of Shakespeare

Scope: The *second law of thermodynamics* occupies a unique place in physics, with applications ranging from everyday experience to the ultimate fate of the cosmos. Unlike other physical laws, the second law is not about what must happen but, rather, about what is unlikely to happen. The essence of the second law is that order inevitably evolves into chaos. The concept of *entropy* quantifies this notion; thus, the second law becomes a statement that entropy generally increases—and, in any event, can never decrease. The second law's consequences for heat and energy are especially significant. The second law makes it impossible for us to extract all the random thermal energy in an object and convert it to more useful forms, such as electrical energy or the energy of motion. That means we can't build perfectly efficient engines, power plants, or other devices that extract energy from heat sources. It also means we can't build a perfect refrigerator—any refrigerator needs an external energy source. Ultimately, the second law is about the *quality* of energy rather than its quantity. So important is the second law that British writer C. P. Snow compared ignorance of the law with not having read a work of Shakespeare!

Outline

I. We're not quite done with physics in the kitchen! Here's a simple culinary act: Break an egg and beat it. Now, if I carefully retrace the steps of the beater, can I once again separate egg and yolk? Of course not! And that's the essence of the *second law of thermodynamics*.

A. There's nothing in the usual laws of physics (Newton's laws of motion, Maxwell's laws of electricity and magnetism) that would prevent the egg from reassembling. It's just extraordinarily unlikely. Of all the possible ways the egg molecules could be arranged, the number of arrangements with all the yolk molecules together is almost infinitesimally tiny compared with arrangements in which white and yolk molecules are intermixed. Given a random beating, then, it's extremely unlikely that the beaten egg will spontaneously separate into yolk and white. How unlikely? So unlikely that if I repeated the experiment every minute for the 15-billion-year age of the Universe, separation of white and yolk would still be extremely unlikely.

B. There are numerous other examples of this concept.

    1. Whoops! I dropped my note cards. If I simply scoop them up, what is the chance that I'll get a coherent lecture? Almost zero. Again, chaos triumphs over order.

    2. I shove a block of wood along the table. It soon comes to a stop because the force of friction (Lecture Eight) opposes its motion. We don't see this, but it also gets warmer as friction converts its ordered motion into the random thermal motion of molecules in the block and the table. A sensitive thermometer stuck in the block would show this, and a movie of the block sliding to a stop and the temperature rising would make perfect sense. But shown in reverse, the movie would look absurd. We never see all the molecules in a stationary object suddenly move in the same direction, giving the object a bulk motion! But that wouldn't violate Newton's laws, or the conservation of energy, or any other "ordinary" law of physics; it's just extremely unlikely.

    3. Put a glass of hot water and a glass of cold water in contact in an insulated container. After awhile, they're both lukewarm, because energy has moved by thermal conduction from the hot to the cold water. Wait a while longer, and they're still lukewarm. You'll never come back to find once again glasses of hot and cold water side by side—although energetically that's possible.

II. All these examples illustrate the second law of thermodynamics. In each case, we begin with an ordered state—separate egg white and yolk; note cards organized into a coherent lecture; a block of wood with its molecules sharing a common motion; and water with faster moving molecules all in one glass, slower moving molecules in another. We end with a less ordered state—a scrambled egg, a random stack of note cards, random thermal motion only, and two glasses of lukewarm water. In all cases there's no going back—at least not spontaneously—simply because the ordered states are so rare, so improbable.

A. The concept of *entropy* distinguishes the ordered from the disordered states. Entropy is a precisely defined mathematical quantity that increases with increasing disorder. The ending state in each example is a state of higher disorder and, thus, higher entropy.

B. In its broadest form, the second law states that the entropy of a closed system can never decrease. A closed system means one that is isolated from its surroundings, with neither matter nor energy flowing into or out of the system. The ultimate closed system is the Universe itself, and in its most cosmic form, the second law thus states that the entropy of the Universe can never decrease. Most processes—such as beating an egg or creating friction—are imperfect in the sense that they result in an irreversible entropy increase.

    1. That closed-system stipulation is important. Locally, entropy can decrease—meaning a system can become more organized. But this can only happen if energy flows into the system. Consider two examples:

    2. Put a glass of lukewarm water in the refrigerator. Eventually, it gets cooler and the refrigerator's external surroundings get a bit warmer, as the refrigerator "pumps" energy out of the water and rejects it to its surroundings (recall the discussion of refrigerators in the previous lecture). This sounds like the situation I just said was impossible, especially if you put another glass of water against the refrigerator's warm exterior coils so that it warms up. Taken in isolation, the system of the two water glasses does indeed get more organized and, therefore, its entropy decreases. But this only happens because the refrigerator is plugged in, using electrical energy. Expand the system to include the refrigerator and the electric power plant that generates the electricity, and you'll find that the entropy of that system increases, as the second law requires.

    3. The appearance of life on Earth, the growth of a plant from randomly distributed molecules of soil and air, my sorting my notes back into a coherent lecture, the appearance of a book from what were randomly distributed molecules of ink, and the development of human civilization—all these are processes that convert random arrangements of matter into highly organized ones. Thus, all reduce entropy—on Earth, that is. But Earth, like the refrigerator, is "plugged in"—in this case, to the Sun, through the steady stream of solar energy that powers life on our planet. Enlarge the system to include the Sun, with the energy-generating nuclear reactions at its core, and you'll find that the entropy of the Earth-Sun system increases. We gain organization on Earth at the expense of more disorganization in the Sun.

C. A narrower but equivalent statement of the second law reads: *It is impossible to build a perfect heat engine.* A heat engine is a device that converts random thermal energy into more useful forms, such as motion or electricity. In this form, the second law says that you can't convert all the thermal energy in a system to useful forms; some of it must remain random. This has profound implications for the use of energy in our technological society. I'll elaborate in the next lecture.

1. The second law is about energy *quality*. Random thermal energy is of lower quality than the energy of directed motion or electricity. You can convert the energy of an object's directed motion to thermal energy, but you can't go the other way with 100 percent efficiency.

2. Even thermal energy comes in different qualities; high temperature represents higher quality. You can extract some useful nonthermal energy from the system consisting of separate quantities of hot and cold water (you'll see how in the next lecture); once they're mixed, that possibility is gone. The greater the temperature difference, the greater the quality of energy (and the lower the entropy). In all these cases, I'm talking about the same *quantity* of energy but different energy *quality*.

D. Another equivalent statement of the second law reads: *Heat won't flow spontaneously from a cooler object to a warmer object*, or more technically: *It is impossible to build a perfect refrigerator.* That's simply a restatement of my example of the two glasses of lukewarm water that won't spontaneously organize themselves into hot and cold—even though there's sufficient energy in the water. A real refrigerator is imperfect, needing a source of additional energy.

1. Back to the kitchen: This is the reason why I said a stove is a much simpler device than a refrigerator; the former easily converts high-quality electrical to low-quality thermal energy with 100 percent efficiency, while the latter has to use a complex mechanism to overcome—only locally!—the second law's prohibition on heat flowing from cooler to hotter. That complicated mechanism, and the energy with which it's supplied, eliminates the word *spontaneously*; therefore, there's no violation of the second law.

   2. Refrigerators and heat engines are inverse devices. More on this in the next lecture.

III. Why the Shakespearean title? And why all this talk of the *second* law of thermodynamics? What happened to the first?

   A. In 1959, C. P. Snow, a British scientist (molecular physics), novelist, and government official, wrote *The Two Cultures*, in which he decried the split between the sciences and the humanities. Snow singled out the second law of thermodynamics as an example of science that every educated person should know; he likened it to a work of Shakespeare—hence my title.

   B. The second law implies there must be a first law. The *first law of thermodynamics* states simply that energy is conserved—although it may be transformed among different forms, including the mechanical energy of bulk motion and random thermal energy. More specifically, the first law says that the internal or thermal energy of an object may be changed either by heat flow or by doing mechanical work. Thus, the first law is about energy *quantity* while the second is about its *quality*. Much more on this in the next lecture!

Suggested Reading:

Louis A. Bloomfield, *How Things Work: The Physics of Everyday Life*, pp. 297–298.

Paul Hewitt, *Conceptual Physics*, chapter 18, especially p. 355 onward.

Going Deeper:

C. P. Snow, *The Two Cultures and the Scientific Revolution*.

Richard Wolfson and Jay M. Pasachoff, *Physics for Scientists and Engineers*, chapter 22, sections 1 and 5.

Questions to Consider:

1. Give two examples from everyday life in which entropy increases.
2. Conservation of energy is a fundamental law of physics. If energy is conserved, why are we so concerned about using it efficiently?

Lecture Twenty-Nine—Transcript
Like a Work of Shakespeare

Welcome to Lecture Twenty-Nine. Let me just check my notes. Oh, yes. This is, "Like a Work of Shakespeare." I've been talking about this lecture for some time, and by the end of it, we will understand why it has that literary title. I'm trying a new system of notes here. Let me begin with a demonstration.

Here's a—we aren't quite done with physics in the kitchen, I guess—here's a glass bowl. Here's an egg. There's the egg in the glass bowl, white and yolk. I'm going to beat the egg. Scrambled egg.

Now, what I would like to do next is take this beater, and very carefully repeat exactly the motions I've performed in beating this egg. I'll turn it the other way—exactly the same speed, exactly the same place—waiting for that yolk to reemerge. Not yet. I'm doing it just right. I'm repeating exactly those motions. It doesn't happen. Why don't you try this at home? Maybe you'll have better luck than I do. Take an egg, and beat the egg with an eggbeater, an electric beater, or a whisk, like I'm doing. Then, carefully repeat the motions, just in reverse, until the white and the yolk separate, and that nice plump yolk reassembles itself out of that mixture of scrambled eggs.

Well, good luck. The chances of that happening in the lifetime of the Universe are minuscule. If you beat an egg every minute from now until the time the Universe has already been in existence, 14 billion years, the chances are negligible that that would happen. Why is that? It's because of all the possible combinations of the molecules that make up an egg, the combination that has all the yellow yolk molecules, and all the albumin white molecules separated, There are many such combinations, but the number of those combinations is far, far tinier than the number of completely random combinations in which egg yolk and egg white are intermixed.

This concept I'm introducing here, with the beating egg, is a probabilistic concept. It's unlike the other laws of physics. There's nothing in Newton's laws that say if I turn the eggbeater backwards at just the right way, the egg will go back together. That doesn't violate any laws of physics, but it is highly improbable that I will ever get the egg package that ordered state.

That is a concept I want to get across now, and let me just check my notes to see what other example—uh! I dropped my notes. What are the chances that when I scoop those notes up, I'm going to be able to give you a coherent lecture. Those notes were nicely in order. They told me exactly what concepts I was going to talk about next, what I was going to see next. Oh dear, they're gone.

If I just go scoop those notes up, I will get some random arrangement of those note cards. Now, there may possibly be other configurations than the one I had them in that made some coherent sense. Maybe there's only that one. Maybe even that one wasn't coherent, but let's suppose that it was. The chances of picking up those notes that fell at random, and getting them into that coherent pattern is absolutely negligible. I could do it one million times, and the chances of them coming out exactly in the right arrangement are negligible.

What am I talking about? I'm talking about a tendency in nature for systems to go from ordered states to disordered states, from your house all neat after you've cleaned up, for it to get messy, for you to drive your kid off to college, come back a few weeks later, and the dorm room looks like a pigsty. We sort of think, "Nature naturally tends to this," and we make some big high-faluting law about it—which I will do shortly.

The real reason that happens is because of all the possible states of the world can get itself into, that the books in your house can arrange themselves into, that might not correct being, the molecules of the egg can be in, the states in which there is clear order—yolks separated from whites, not cards in the order that gives a coherent lecture, all the books on the shelves in nice alphabetical order—those states are far fewer than the more random, chaotic states; and so, left to its own devices, if change occurs at all, it will generally, just out of this probability that there are more random, unorganized states, things will evolve toward this more unorganized state. It is conceivable to imagine that they would go the other way, but it just doesn't happen, and the reason it doesn't happen is because there are so many more chaotic states, so many more disorganized states, than there are organized states.

Let me give you one other example of that. Here I have one of these blocks that I used to demonstrate friction a long time ago, in Module Two. Friction, remember, in a non-conservative force. Mechanical energy, the bulk energy of motion, is not conserved in friction. Something is conserved in friction, and we will get to that soon.

Let me just give this blocks a shove along the table, and it quickly comes to a stop. Now, where did its energy go? Its bulk mechanical energy is gone, but energy is conserved, and we know where that energy has gone. The table surface and the block have become a bit warmer. That bulk translational of the entire block moving has been converted into that random motion of the molecules we call, loosely, "heat," or that we should more correctly call "internal energy."

In fact, if I had stuck this thermometer into the block as I did this experiment, this wouldn't really have happened, because the energy involved in relation to the energy needed to raise the temperature in this block, significantly, is small; but in principle, if I shoved this block along, I would see the thermometer go up.

Because that's not something we can actually do, I have made a little movie of that here. Here I have the block, with the thermometer stuck in it. The block is going to startup moving rapidly, and is going to decelerate, because the force of friction is acting on it. It is going to give its mechanical energy, and that energy is going to reappear as the internal energy, the random thermal energy of its molecules. As a result, it's going to get hotter.

This is what the movie looks like. The block is moving along fast at first, it's slowing down, you see the temperature going up, and up, and up. The block is going more and more slowly, because it is decelerating, and eventually, it comes to a stop. There, it is slowing and slowing, and eventually, it comes to a stop at a much higher temperature, or a somewhat higher temperature.

That is a reasonable movie. That is something that you are entirely likely to see in the real world. You might not actually be able to stick the thermometer in, but a block going along and coming to a stop is not the least bit surprising.

Here's what's very surprising. That block is sitting there, and it has all of that random thermal energy in it. What if all of those molecules decided, all at once, to be moving in the same direction, so that the block suddenly took off, and started heading in that direction all by itself? You might say, "That's impossible." It's not impossible, because the energy isn't there. The energy is there. The energy is there in the random thermal motions in all directions, and it could happen that they just conspired instantaneously to all go in the right direction, and then the block would start moving.

That would not violate the conservation of energy, but it would be enormously improbable. Here's another movie. Here's the block where we left it. Now, the block starts out slowly at first, and gradually gains speed, moving to the left. As it does so, the temperature drops, and it goes faster and faster to the left. There is nothing in that movie that violates conservation of energy, and yet, you would say that something was wrong with that movie, that I'm playing the move backwards, or we are running the projector the wrong way, or something.

It doesn't happen. Why doesn't it happen? Because of all the possible distributions of motions, the particles in that block could have, the one in which they're all moving in the same direction is very, very unusual. If I am ever in that situation, as I am what I give the block a big push, the chances are, as that system evolved some other configuration, it will evolve to one of those much more likely configurations in which the motions of the molecules are not all directed in the same way, but are random.

In all of these examples—the beating of the egg, the dropping of the lecture notes, or the wooden block—the reason we tend toward a system of more disorganization is because those states of motion are more probable. Let me give you one other example, more closely related to the topic of this module, namely heat, and thermal things.

I'm going to move over to this microwave oven, where I have a glass of cold water sitting on top, and a glass of hot water that has been sitting in the oven. I take them, and I put them side by side. I'm just going to let them sit there for a while. We both know what's going to happen. I'm going to come back, and I'm going to have two glasses of lukewarm water. The cool water is going to have warmed up to room temperature, and the warm water is going to have cooled down.

By the way, even if I had put them in a completely insulated container, a great big Styrofoam cooler chest or something, they would exchange energy with each other. Heat would flow from the hotter glass to the cooler class, and I would come back and would find them lukewarm. I could wait from now until forever, and it would be very unlikely that the energy would flow back from one lukewarm glass to the other one, raising the temperature of one, and cooling the other, although the total energy would still stay the same.

We are talking here then in this lecture, about something that is a little bit different than energy and its conservation. We are talking about a new law, physics that is different than the laws of Newton, which are absolutely strict, and must be true. It's a probabilistic law. It says that things are unlikely to go spontaneously into more organized states, simply because there are more disorganized states available to them.

All of these four examples I've given you now—the egg, the lecture notes, the block, and the water—all illustrate that law, which is called the "second law of thermodynamics." The second law simply says that given their own devices, systems move naturally toward states of less organization and more chaos. To be a little more strict, the second law of thermodynamics says that the best you can possibly do is to stay in a particular ordered state. You can't get more organized spontaneously. I'm going to spend the rest of this lecture elaborating on that concept, the second law of thermodynamics, and the related concept, which you have probably heard and sort of know the meaning of, the concept of *entropy*.

Entropy is a quantitative measure—we can actually calculate a numerical value for entropy—that describes how ordered or disordered a state is. Thus, the entropy of this scrambled egg here—not cooked yet, but scrambled—that entropy is greater than the entropy of the egg when I first broke it, and it had separate white and yolk. Entropy is a measure of disorder; so, the second law of thermodynamics basically says that entropy tends to increase. More specifically, it says that the entropy of a "closed system"—I'm going to talk more about closed systems in a minute. What I mean by a "closed system" is a system that doesn't have any communication with the rest of the world, or the rest of the Universe. It's completely closed; in particular, it can't exchange energy with the outside world, or with anything else, for that matter.

The second law of thermodynamics says that if you have a closed system, the best you can do with that system is to keep its entropy constant, to keep its state of organization unchanged; but more generally, what is most likely to happen is that it is likely to become more disorganized, and entropy is likely to increase. The second law, put more succinctly, says that the entropy of a closed system cannot decrease. It might just barely be able to stay the same, and is more likely to increase, but not decrease. If you want a simple English translation of the second law, the English translation is, crudely, that states evolve naturally toward more chaotic situations, toward more chaotic organizations. Systems evolve toward states that are more chaotic, and less organized, with more entropy. If you want to be really precise, they either stay as they are, or they evolve toward more disorder, but they don't evolve spontaneously and naturally toward states that have more order.

What's a system? Well, that's your choice. The room I'm in right now could be a system. My body could be a system. Planet Earth could be a system. You choose the system. The second law applies to closed systems,

systems that cannot communicate with the rest of the world. Well, all of those systems do. I'm certainly interacting with the rest of the world. This room that I am in is interacting with the rest of the building, although because this room is air-conditioned, because of the bright studio lights, they've taken precautions to minimize that interaction so that they can keep this room and its surroundings cooler, for example, so we can achieve with insulation and things like that some degree of isolation with the rest of the world, but we really can't get completely there.

We therefore might expand our definition of systems until we get to the entire Universe, because that's all there is; and so, there is no question of the Universe being isolated from everything else. Well, we don't want to go there, but there are multiple Universe theories and things like that, that we can get to in exotic theoretical physics courses, but not in here. Let's stop at the Universe. The second law then simply says that the entropy of the Universe cannot decrease. Period.

The implication of that statement is that it can increase, and although the statement doesn't imply it, it is, in fact, likely to increase. Once it does increase, it can't go back. Therefore, entropy at best stays the same, and more likely increases. That is the statement of the second law, and most of the processes that we tend to undergo, processes that involve non-conservative forces like friction, processes like the beating of the egg, processes like putting the two glasses of water that were once hot and cold—a kind of organization there, together, and have them become lukewarm—those are imperfect processes in the sense that they're processes that result in an increase of entropy, and a loss of order.

Now, I could do something interesting here. I could walk over to these glasses, which are now the same temperature, and I could put one of them in the refrigerator. I could put the other one against those coils that I talked about in the last lecture, where the refrigerator rejects its waste heat, and I could run the refrigerator a while. The one lukewarm glass would get cooler, the other lukewarm glass would get warmer, and I would, in fact, have transferred energy from the cooler glass to the warmer glass. When it was all over, I could put the glasses back together, and I would have that nice, wonderful, organizational state I started with. That would be akin to reversing the egg beating process, and separating the white and yolk, so why doesn't that work? Why can't I do that?

Well, I can do that. I can do the experiment I just described. I can put those glasses in the refrigerator—one in the refrigerator, one outside—and do that experiment. Here is a conceptual picture of what that experiment looks like. Here I have one glass in the refrigerator. The other glasses outside the refrigerator, down near those coils where the refrigerator rejects its waste heat. They're both of the same temperature, as indicated by the thermometer. After I run the refrigerator for a while, energy is transferred through that process by which the refrigerator works from the glass inside the refrigerator to the glass of water outside of the refrigerator. Water in the refrigerator becomes cooler, and the water outside becomes warmer. We have that organized state again.

Okay? There has been in entropy decrease. There has been an increase in organization. We have moved from a chaotic, more randomness, to a state where there's order, organization. Glass A is warm, and glass B is cool.

If I draw a box around the refrigerator, and the glass that is outside, I have a system. That is a system where entropy has decreased. Does that violate the second law of thermodynamics? No, it doesn't, because that is not a closed system. There is something else involved here. There is a wire connecting the refrigerator to a power plant somewhere, so the refrigerator is not a closed system. Maybe or maybe not, but if I draw a box around the whole system of power plant and refrigerator, what the power plant is doing is burning fuels, and all kinds of processes are going on in the power plant that ultimately increase the energy. The chemical composition of a fuel is more organization than the chemical composition of the products of combustion, for example.

The process at the power plant, then, is to increase entropy. The process at the refrigerator is to decrease entropy, but the refrigerator cannot be a closed system. This is why it has to be plugged in. The second law says so. That is why it is a profoundly different kind of device, as I suggested last time, then a simple stove or heating device. It has to be plugged into the power plant, and if I look at the entropy of this entire system—the increase of the entropy at the power plant, the entropy decrease at the refrigerator— there is an overall entropy increase. The refrigerator, therefore, has not solved the problem, and has not violated the entropy law of the second law of thermodynamics.

Now, let me give you some other examples. Life appeared on Earth. We had the planet, initially, kind of in a state of chaos, and life has evolved on Earth. We humans have evolved, and we have done things like put symbols on printed pages, or, in the case of the course you are watching, probably, little bit in a very carefully organized pattern on a little plastic disk, DVD, or, if you are old-fashioned, watching on VHS tape. We have imposed a definite information-containing organized magnetization pattern on a strip of VHS tape, where before, it was always random.

Organization has appeared out of chaos on planet Earth. All of these processes reduce entropy. Evolution—the evolution of life, and of intelligence, and of human society—these are entropy-decreasing processes. How is that possible?

It's only possible because Earth is not a closed system. Earth is just like the refrigerator. It's connected to something. In the case of the refrigerator, there's a wire connecting the refrigerator to a power plant. In the case of the Earth, there's a steady stream of electromagnetic radiation coming from the Sun. There are entropy-increasing processes occurring in the nuclear fusion reactions that occur in the sun's core, and the combined system of Earth and Sun is a system in which entropy is increasing. Entropy is decreasing on Earth, as we organize things, and entropy is increasing on the Sun, due to the nuclear reactions at its core. Additionally, the overall system is showing an entropy increase.

There's nothing wrong with making organization happen. You are allowed to clean up your kid's dorm room, your room, or vacuum your house, or whatever, and it will go to a more organized state, but I guarantee that energy will be used in the process, and the processes that produce that energy will themselves generate more entropy, and the overall effect will be to increase the entropy of the Universe. You gain organization, but it's at the expense of disorganization, possibly, somewhere else.

Now, let me give you another and more specific, and a sort of engineering-oriented, statement of the second law of thermodynamics, which will play a major role in the next lecture, when we talk about us as high-energy users, something that I hinted at a long time ago, when I had Jamiee up here cranking that generator.

One statement of the second law is that it is impossible to build a perfect heat engine. What is a perfect heat engine? Well, my example of the block that started up, and somehow converted all of that random thermal energy into directed motion is sort of a heat engine. Here's a conceptual diagram of what a perfect heat engine would be. This isn't a real engine. An engine is a device that produces mechanical energy out of something, typically by burning a fuel. We will conceive that somehow, we've gotten some stuff hot—water; I don't care what—burning a fuel, making a nuclear reaction, using sunlight, I don't care, but we have a hot reservoir. It's a bunch of stuff at a high temperature.

Conceptually, what a heat engine would do is that it would be a device that would somehow tap the internal energy of his heat reservoir and produce useful energy—mechanical or electrical energy coming out. We would extract all of that random thermal energy as useful work.

That's what the second law says is impossible. The second law says that if we want to build a heat engine, we have to do something like this, and I will describe an implementation of this in the next lecture. We have to have a cold reservoir; cooler than the hot one, and the cool reservoir could just be the ambient environment around us. It usually is, in the case of the engines we build. What the second law says is that if you try to extract heat from that reservoir, what you will do is to extract energy. Some of it will turn into useful work, but you can't turn all of it into useful work.

Some of it—in fact much of it, in the case of the engines we build, two thirds of it, in the case of electric power plant, typically—gets rejected to the environment as waste heat. There is no way around that. That's a fundamental limitation posed by the second law of thermodynamics. One consequence of the second law of thermodynamics then is that is impossible to build a perfect heat engine. You can try to extract energy from thermal sources, and turn it into useful mechanical work, and maybe run an electric generator, or drive the wheels of a car, but in the process, the second law guarantees that you are going to throw a lot of energy away as waste heat.

Now, why do I call it "waste heat?" It is useful. In fact, you could use it to heat your house. In fact, there are cities in Europe, particularly, where houses are, in fact, heated by that waste heat from power plants. There are industries, increasingly, in the United States—more about this in the next lecture—that need hot stuff, hot water, steam, and electricity, and they generate them in the same system, and they use the waste heat as the usable heat that they need, so that it isn't totally wasted.

What we are talking about here, really, is energy quality. What the second law is about is telling is not about amounts of energy; remember, that there was nothing wrong with that movie running in reverse energetically, and nothing wrong with any of these things, as I said, in that they can happen energetically, it's just that they are improbable.

With the second law is about is energy quality. Here's a diagram that tells you a little bit about the quality of energy. The best kind of energy that you can have, the highest-quality energy, is mechanical or electrical energy. You can do anything that you might want to do with energy, with other mechanical energy, the energy of direct bulk motion of an object, the rotational motion of a machine's shaft, the motion of a car, the motion of an airplane through the air. These are the highest-quality energy, or electrical energy. Mechanical and electrical energy are of the highest quality, and the lowest entropy. Heat at very high temperatures, because we are in environment at low temperature, represents a great deal of organization, and therefore, high quality, and low entropy. The lowest temperature represents the lowest quality, so on this scale, energy quality goes from low to high on the left there, and entropy goes from low to high, downward.

There are different values of energy quality. You can convert the highest-quality energy to the lowest with 100% efficiency. It's a dumb thing to do, because you are wasting its good quality. That's why an electric water heater, for example, or electric heating in general, is not a great idea, because it just takes low quality energy to heat your house, if you've got high quality energy, why not do something really useful with that, rather than turn it into low quality heat? You can do that though, and you can do it with 100% efficiency.

If you have low quality energy though, you can't turn it back to high quality with 100% efficiency. That's why this block, it's full of low quality energy, enough to make the block move like this, but it won't happen. I could build a machine that would get the block moving, but most of the energy it extracted from the block would end up as waste heat, rather than as the motion of the block. That's the second law of thermodynamics, talking about energy quality.

There's another equivalent statement, and the equivalent statement says that it is impossible to build a perfect refrigerator. That's another statement; and you can prove—you can do this if you took a standard introductory physics course—that those two statements of the second law are logically equivalent. If one of them is true, the other one has to be true.

What is a perfect refrigerator? Well, here's a perfect refrigerator conceptually. What a perfect refrigerator does is to say, "Okay, I want to extract some energy from some cold stuff, and I want to put it out in some hot stuff. I want to take as much energy as I can out of the cold stuff, and have all of that energy flow into the hot stuff; I don't want to do anything else." That would be the spontaneous flow of energy from a cool thing to a hot thing. That would be my two glasses of lukewarm water; one of them getting cooler, and the other one getting hotter spontaneously, energy flowing from the one was cooling to the one that was heating. The creation of order out of what had been a more chaotic, more disorganized state. The second law says that you can't do that.

How do you build a refrigerator? Well, I showed you. Unlike with an engine, I showed you the actual mechanics of how you would build a refrigerator. It doesn't look like this picture on your screen, but it's conceptually like that. However, it's missing the part that makes it a real refrigerator. Think about what that part was. It was the part where I plugged the refrigerator into the wall, ultimately into some power plant, which has some kind of nuclear, chemical, or whatever kind of reaction that's going on to increase entropy at the power plant.

The real refrigerator looks like this. It's the opposite of an engine. An engine is a device that takes heat from a hot reservoir, produces useful mechanical energy—high quality energy, mechanical energy or electrical energy—and rejects some waste heat to the environment. A perfect refrigerator goes the other way, or a real refrigerator goes the other way. It extracts energy from a cool reservoir and dumps it into a hot reservoir, but it can't do that spontaneously, completely. It requires an input of useful, high quality energy, mechanical or electrical energy. That's why you've got to plug your refrigerator in.

By the way, that also makes heat pumps for heating houses not perfect devices, but they're a lot better than heating your houses with, say, electrical resistance, and they may be even better than burning fuel, because what heat pump does is to use some useful energy, some high quality energy like electricity, to run what is basically a refrigerator. It refrigerates the ground, or, if you're in a southern state, the outside air, cools it down a bit, and dumps that heat into your house. If the heat pump is well-designed, and the temperature differences are right, it may take, for instance, one unit of mechanical energy or electrical energy, to move three or four units of heat. In this case, you've gotten a lot more energy into your house and you brought in the form of electrical energy, so that those devices can be quite helpful, and quite efficient.

This, by the way, brings back the point I made in the previous lecture. Again, I said the refrigerator and stoves were very, very different kind of devices. Refrigerators are devices that involve the second law of thermodynamics and its fundamental limitation, because they have to have high quality energy coming in, but when you just want to heat water, you can get away with the lowest quality energy, and you are fine.

Well, that almost brings me to the end of this lecture, but I wanted to give you a reason, justification, for the title, which I have been talking about since the beginning of this course: "Like a Work of Shakespeare." There was a British writer and scientist, and he was also a government official, named C.P. Snow. He wrote a book called *The Two Cultures*. He decried the fact that the sciences and the humanities could never talk to each other.

There's a famous quote from C.P. Snow, and in that longer quote is the title of this lecture. I want to share that with you. This is why the title is, "Like a Work of Shakespeare." C.P. Snow said, "A good many times I have been present at gatherings of people, who, by the standards of the judicial culture, are thought highly educated, and with considerable gusto have been expressing their incredulity at the illiteracy of scientists. Once or twice, I have been provoked, and I have asked the company how many of them could describe the second law of thermodynamics."

Let me parenthesize here for a moment. For what I said in this lecture, I think you can see that the second law of thermodynamics is pretty basic, and pretty important, particularly to we energy-consuming denizens of the planet Earth, and if that's not obvious now, it will become much more obvious in the next lecture.

C. P. Snow went on, "The response was cold. It was also negative. Yet, I was asking something, which is about the scientific equivalent of, 'Have you read a work of Shakespeare's?'" That's why my title is, "Like a Work of Shakespeare." The second law of thermodynamics is like a work of Shakespeare. It's something that ought to be in the repertoire of every intelligent educated person, like a work of Shakespeare, like the second law of thermodynamics.

Now, before I end, there's something about this second law of thermodynamics that implies that perhaps there's a first law of thermodynamics. Well, there is a first law of thermodynamics. The first law of thermodynamics is basically a broadening of the concept of energy conservation. I already talk about the conservation of energy, and how we could swing from potential energy to kinetic energy, and back and forth, or bounce on a spring, and store energy in the spring, and so on. Then, though, we worried about non-conservative processes like friction. Where do they fit in?

Well, the first law speaks to them, and it says that basically, energy is conserved, except that that includes, now, internal energy. It says that you can do useful work on something, and you can raise its internal energy, or you can transfer internal energy by heat, but however you do it, energy is ultimately conserved; however, it can be transferred among forms, including these lower quality forms, like internal energy.

The first law, then, is about the conservation of energy as a quantity, whereas the second law speaks to something entirely new and different, namely, energy's quality.

Lecture Thirty
Energy in Your Life

Scope: Modern humanity uses energy at a prodigious rate. The human body produces energy at the rate of about 100 watts, the equivalent of a 100-watt lightbulb. The average citizen of the United States uses energy at the rate of about 100 human bodies' worth—as if we had 100 "energy servants" working for us round the clock. Most of that energy comes from burning fossil fuels; much of the rest is from nuclear sources and from the burning of garbage, wood, and other biomass. All these involve thermal energy and are subject to the laws of thermodynamics. Some of our energy use is of the highest quality, or lowest entropy. This includes energy used in transportation and electrical energy. Other energy uses call for lower quality energy, as in space heating, water heating, and some industrial processes. The second law of thermodynamics puts stringent limits on our ability to extract high-quality energy from thermal sources. As a result, our electric power plants are typically only 30 to 40 percent efficient, and our cars and trucks, only about 25 percent efficient. We could use energy more efficiently if we paid increased attention to the concepts of entropy and energy quality.

Outline

I. Modern humanity uses energy at a much greater rate than our bodies can provide. The hand-cranked generator shows that the human body can produce energy at a sustained rate of about 100 watts. Deep knee bends at about once a second produce energy at about the same rate.

  A. We need to distinguish energy and power. *Energy* is the actual amount of "stuff," while *power* is the rate at which it's used or produced. Common units of energy include joules, calories, British thermal units, and kilowatt-hours. Common power units include watts and kilowatts, horsepower, and Btu/hour. Energy and power are often confused. To say that the average power output of the human body is 100 watts is to give the rate at which the body produces energy. The total amount it produces depends on how long it's producing at this rate.

B. The average U.S. citizen of the early 21st century uses energy at the rate of about 10,000 watts—10 kilowatts (kW), or 100 times the body's rate of energy output. That's like having 100 "energy servants" working for each of us round the clock. (The number is less in most but not all other countries; in Western Europe, it's about 50 "energy servants.")

C. Those energy servants perform a lot of useful services, including, broadly, transportation, industry, and residential and commercial energy supply.

D. Most of that energy comes from burning fossil fuels. Much of the rest is from other thermal processes, including nuclear fission and the burning of garbage or biomass.

II. To extract useful, high-quality energy from thermal sources, we use *heat engines*. Examples include the gasoline and diesel engines that power our vehicles, jet engines used in aircraft, old-fashioned steam engines, and the steam boilers and turbines used in electric power plants.

    A. Conceptually, all heat engines operate in much the same way, going through a sequence of steps that typically involve expansion, compression, and/or state change of a fluid.

        1. Heat flows into the fluid, which expands, pushing on a piston and producing high-quality mechanical energy.

        2. The heat flow stops and the fluid continues to expand, now cooling. More mechanical energy is produced.

        3. To get the fluid back to its original state, it's cooled and compressed. During this phase, heat flows from the fluid to the environment; thus, some of the energy that originally entered the fluid is lost as waste heat, rather than being turned into mechanical motion.

        4. The fluid is isolated from the environment and continues to compress to its original state. The cycle then repeats.

    B. Although real heat engines use different processes, they all share the common phases of heating, expansion, cooling, and compression—and all involve turning some but not all of the energy from a thermal source into high-quality mechanical energy.

        1. Examples include gasoline and diesel engines, in which heating and expansion are associated with fuel burning in a cylinder, and steam power plants, in which heating by fossil or nuclear sources turns water to steam, which then drives a turbine. A car's radiator and a power plant's cooling towers all emphasize the fact that some energy is lost as waste heat.

2. Couldn't a clever engineer design a more efficient heat engine that didn't have to waste any energy? No! In 1824, the French physicist Sadi Carnot proved that no heat engine can be more efficient than the *Carnot engine* that I described above and that the efficiency of this engine is limited by the ratio of the minimum to maximum temperatures involved in the engine. The greater the extremes of temperature, the more efficient the engine. An engine with 100 percent efficiency would require either an infinite high temperature or absolute zero low temperature—both of which are impossible.

3. We're stuck with Earth's ambient temperature at the low end, and the strengths of typical materials limit high temperatures, with the result that typical heat engines have efficiencies in the 25–40 percent range; even the most advanced barely exceed 50 percent. That means we throw away somewhere between half and three-fourths of the energy we extract from fuels.

4. That isn't just waste but may also cause *thermal pollution.* For example, a substantial fraction of the total river flow in the United States makes its way through the cooling systems of electric power plants, raising water temperatures and potentially affecting river ecology.

5. Couldn't we do something with that waste heat? Yes! Stay tuned.

III. Energy uses can be distinguished by the quality of energy that's needed.

    A. Some energy uses require relatively low-quality energy. These include space heating, water heating, and some industrial processes.

        1. Such energy can be produced directly by burning of fuels, as in a home heating system, water heater, or industrial boiler. About two-thirds of the energy need in the United States is for such low-quality energy.

        2. Most of the energy produced in burning can be captured and used to heat water, air, or other substances. Thus, direct burning is fairly efficient and could supply our low-quality

energy needs. The second law of thermodynamics places no restriction on such energy conversion.

B. Transportation and electrical energy for motors, lights, and electronic equipment require high-quality (low-entropy) energy. These uses make up about one-third of our energy needs. Again, the second law of thermodynamics says we can't convert from thermal energy to high-quality energy of motion or electricity with 100 percent efficiency.

C. Although we need high-quality energy for only about one-third of our energy uses, about two-thirds of the energy we produce is of the highest quality. Therein lies huge inefficiency and energy waste.

    1. Consider domestic water heating. Some homes burn gas or oil to heat water, either in a separate heater or as part of the home's space heating system. But many use electric water heaters. A good gas heater recovers about 80 percent of the fuel energy to use in heating the water; for an electric heater, the figure may be as high as 95 percent of the incoming electrical energy ending up as thermal energy in the water. But which is really more efficient? Pause and think about this! Assuming the electricity is produced in a power plant with a typical efficiency of only about 35 percent, the electric heater is much less efficient.

    2. All that waste heat from our power plants could be used to supply low-quality energy needs. That's done in some countries, where waste heat from central power plants has been piped to the surrounding communities for heating. More recently, energy deregulation has favored the practice of *cogeneration*, whereby industries and other large institutions generate their own electricity and simultaneously use the waste heat, often in the form of steam, for processes requiring lower quality energy. Europe, the world leader in cogeneration, is approaching 30 percent cogenerated electrical energy.

IV. Another approach to more efficient energy use is to employ *heat pumps*, of which refrigerators and air conditioners are examples. With these, we're generally concerned with cooling, but conceptually, a refrigerator or air conditioner is simply a device that pumps heat energy from cooler objects (the refrigerator contents; a room or house) to a warmer ambient environment (recall the previous lecture).

A. Run an air conditioner in reverse and you have a heat pump, extracting heat from the cooler outdoor environment in the winter and pumping it into a house. In the southern United States, the same device is used for cooling in the summer and heating in the winter, exchanging heat between the house interior and the outside air. In northern climes, it's more efficient to exchange heat with the ground, through a system of buried pipes.

B. A heat pump is, conceptually, a heat engine going in reverse. An engine extracts energy from a hot source and delivers high-quality mechanical energy as well as waste heat rejected to a lower temperature environment. A heat pump takes in both energy from a low-temperature environment and high-quality energy (for example, electricity used to run a refrigerator's compressor) and delivers energy to a higher temperature environment.

    1. Applied to heat pumps, the second law of thermodynamics says that the greater the temperature difference between low- and high-temperature regions, the more high-quality energy must be supplied.

    2. For example, a heat pump extracting energy from the ground at 50 °F and delivering it to a water heater at 150 °F requires, in theory, just one unit of electrical energy for every six units of energy supplied to the water. Thus, a heat pump can be much more efficient than direct use of electricity or even than burning of a fuel.

V. The examples in this lecture represent some ways to reduce the huge number of "energy servants" at our beck and call. All are based on understanding the important limitations posed by the second law of thermodynamics. But these are only a few of the many ways we might reduce our use of energy and/or increase the efficiency with which we use energy. Could we get away with 50 energy servants instead of 100? There's enough in that question for another whole course!

Suggested Reading:

Louis A. Bloomfield, *How Things Work: The Physics of Everyday Life*, chapter 7, section 1, especially pp. 299–302.

Paul Hewitt, *Conceptual Physics*, chapter 18, especially pp. 349–352.

Going Deeper:

Richard Wolfson and Jay M. Pasachoff. *Physics for Scientists and Engineers*, chapter 22, sections 2 and 3.

Questions to Consider:

1. On a cold morning, some folks will turn on the oven and leave the door open to help warm the house. Not a very safe practice, but it works. On a hot day, could you cool the house by leaving the refrigerator door open? Explain your answer.

2. In the heat-pump example in this lecture, you use one unit of electrical energy and end up with six units of energy supplied to heat water. Where did the extra energy come from?

Lecture Thirty—Transcript
Energy in Your Life

Lecture Thirty, "Energy in Your Life." If you think back on the other lectures you've had already in this course, many, many of them deal with the theme of energy. We ended the previous lecture on the subject of the first law of thermodynamics, an extension of the idea of energy conservation, from mechanical energy, potential and kinetic, to the more general sense of energy, including this internal energy that we associate loosely with the word "heat," and energy that can't be easily extracted and turned into mechanical energy or electricity.

We talked about the energy of orbiting satellites. We talked about the kinetic energy of motion, in the module on going places. Energy has been a major theme in this course; and in this lecture, I want to focus on energy in your life, particularly, the energy that you use as a citizen of industrialized society in the 21st century. That's the theme of this particular lecture.

The fact is that we modern humans use energy at a far greater rate than our own bodies can supply. That's the big idea and the big theme here. I want to give you a couple of examples that demonstrate that.

Some lectures ago when we were talking about electromagnetic induction, I had Jamie from The Teaching Company up here, and I was kind of tormenting her with this device, which is a hand-cranked electric generator. We read the voltage in the current on these meters, and if we multiplied them together, we got the electric power we were producing. I illustrated the generator then largely to show you how electromagnetic induction was used to produce electricity. Now, I'm using the generator as symbolic of the energy that my body can produce, and I want to get you a real gut feel for that energy.

I tried to do this before, but I'm going to emphasize it now more in the context of how much energy you use as a citizen of industrial society, versus how much your own body can produce. Your body is a source of energy. You eat food, you metabolize the food, and you can move, you can lift things, you can run, you can swim, you can jump, you can dance, and you can do all kinds of things that involve turning that food energy into mechanical energy, basically. You also think, and your blood courses through your veins; all kinds of things happen.

How much energy is your body capable of producing, versus the amount of energy you use as a resident of industrial society? I think I won't torment any more Teaching Company employees. I will just crank away on this thing myself; and again, it's very easy to crank it, provided that there's no demand for electrical energy. I'm not producing any energy, except for a little bit to overcome the friction in the shafts of this device; but when I turn on this hundred-watt light bulb—uh! I can barely turn it. It's all I can do to make that hundred-watt light bulb light, and I'm succeeding. It's pretty much at full brightness. I've got 12 volts, and somewhere between half and one ampere [sic 5 and 10 amperes]. I'm succeeding right now in producing energy at the rate of 100 watts.

Now, I probably don't want to do this all the time. On the other hand, it's not something I can only do for a second or two. I can probably comfortably produce energy at this rate for a while. It depends on how buff you are whether you could do this longer, or if you could light a bigger light bulb; but that light bulb, at 100 watts, is pretty typical of the human body. The power output of the typical human body is, in fact, about 100 watts.

Though I might not want to do that day in and day out, there are other processes going on in my body, for example, the keeping of my body at 98.6 degrees Fahrenheit. There is an inevitable heat loss to the cooler environment, and consequently, my body is always producing energy. In fact that figure of 100 watts, although it looks like it takes a lot of work, is, in fact, what my body is producing all the time.

If you don't have access to a hand-cranked generator—if you do, try it out. Really, it's good to feel what 100 watts means. I don't just want you to know some complicated mathematical formula for power. That's what 100 watts means.

If you haven't got that kind of thing, do some deep knee bends at perhaps the rate of about one per second, going down, and up, and down, and up, and down and up, at the rate of about once per second. Now, it depends on your mass—how big you are—and how far down you are going, and how fast you are doing them, but if you are a sort of normal sized person, chances are that the power involved in doing that—you could calculate this using principles of first-year physics—is about 100 watts. The point is that a human being is roughly equivalent to a 100-watt source of energy—give or take a little, but roughly 100 watts.

Now, at this point, I want to take a minute and be sure to make a distinction that I've made a couple of times before, but we really have to get it down solidly here. That's the distinction between energy and power. Energy is a kind of "stuff," as I've argued, one of the basic substances of the Universe, what makes everything happen: Kinetic energy, potential energy, internal energy, heat—energy flowing because of a temperature difference.

Power is a measure of rate. In the same way that miles per hour is the rate at which your car covers ground, or eats up miles, if you want to put it that way, or produces distance, similarly, power, measured in watts, is a measure of the rate at which a system produces—or consumes, transfers, or whatever—energy.

Some common units for energy that you may be familiar with—I've been talking about *joules*. The joule is the unit of energy for the official international system of units. A joule is the energy that you would produce if you exerted a force of one newton for a distance of one meter. That's the definition of a joule.

A calorie is the amount of energy. It used to be thought it was a unit of heat, but then it was discovered that heat and thermal energy or mechanical energy were basically the same kind of thing, the mechanical equivalent of heat—an experiment that Joule did proved that. That's why he's honored with that name. "Calorie" is the amount of energy that it takes to raise the temperature of one gram of water one degree Celsius.

A British thermal unit—used in the United States—we tend to think of these as heat energy, but I could cut about the amount of British thermal units that I supply to a bowling ball by lifting it up, or something. A British thermal unit is the amount of energy that it takes to raise one pound of water one degree Fahrenheit—English units.

Another unit of energy—this is the one you see on your electric bill—is the *kilowatt-hour*. A "kilowatt-hour" is the amount of energy you would use if you ran a 1,000-watt light bulb—well, that's a pretty hard light bulb to get hold of—or a 1000-watt hairdryer for one hour. That would use a kilowatt-hour. If you ran a 100-watt light bulb for 10 hours, that would be a kilowatt-hour. By the way, you probably pay about $0.10 per kilowatt-hour. That means that if you were paying me to produce that kilowatt-hour for you, turning this crank, and you were doing so at rates competitive with your electric company charges—that's why I said a tenth—you would pay me one tenth of one cent per hour. Electricity is truly cheap. If we had to produce our electricity with human energy servants—more on that in a minute—turning cranks, would be paying them well below minimum wage, way under one penny per hour.

Notice kilowatt-hours. That's kind of a funny energy unit. That's because the watt, or the kilowatt is a rate of power, a rate at which you are using energy, so that when you multiply by time, you get a unit of energy. In the case of watts then, unlike miles per hour, watt sort of has the "per time" built in. A watt is a joule per second. A kilowatt is 1,000 watts, 1,000 joules per second. We know in our guts—not mathematically, but in our guts, from this demonstration—what a watt is. There's 100 watts. It's that, "Uh!" as I'm turning that crank, or the "Uh!" as you are doing those deep knee bends. That's what 100 watts means.

Energy and power often get confused. You'll read in the newspaper that a new power plant is going to produce 1,000 megawatts every hour. Well, that's nonsense. "Watts" already has a time built into it. It produces 1,000 megawatts, or a gigawatt. That's the rate at which it produces electrical energy.

To say that the average power of the human body is 100 watts is to say that that's the rate at which the human body can transform food energy and other forms of energy, produce useful energy. If you ask, "How much energy does my body produce?" that's a very different question. That depends on how long I choose to crank that crank, or how long I live, or whatever. The power is the rate at which I can produce energy. It doesn't tell you anything about how much energy I produce, unless you also know how long I'm putting out that power, or how long this light bulb is using that power, or whatever.

Now, here's the question I want to ask. At what rate do we citizens of the 21st-century United States, 21st-century industrialized society, use energy? Around the clock, on average, in our names, at what rate is energy being used in our names?

Before we answer that question, think for a minute about it. Well, there are lights. The lights are burning to make it right for you after the natural daylight is gone. The air-conditioner is running to keep you cool, or the heat is running to keep you warm. You are burning fuels, and running that electric blower that blows the warm air around her house, and that refrigerator is running. As we know from the previous lecture, it requires electricity. You wash your clothes with a washing machine that has a big motor, and uses hot water that was heated with some energy source, and you dry your clothes with a clothes dryer—you don't have to; you can hang them on a clothesline—but you do probably dry them in a clothes dryer.

You drive a car, and maybe it's a big SUV that consumes an enormous amount of energy. Maybe it's a small car, but it still consumes a lot more energy than your body could produce; unless you ride a bicycle or walk, that's the case. Maybe you fly around in airplanes. You ride in trains. You go to the supermarket, and there are banks, and banks, and banks of frozen foods, and so that you don't have to be inconvenienced by opening the door to get to those, some supermarkets are still having those open banks. There's an enormous amount of energy going into cooling that food, because it is exposed to the warmed air of the supermarket. What a colossal waste.

There is a large amount of energy being used in your name. How much? Well, I'm going to give you the answer, although if you have thought about it awhile, you might be able to come up with the answer. The answer is this: For Americans—and I'm speaking, now, particularly, to Americans, citizens of the United States, particularly; Canadians are nearly the same, actually, a little higher on this number—Americans of the late 20th and the early 21st centuries consume energy at the average rate of a little over 10 kilowatts, 10,000 watts. That's the rate at which energy is being consumed in your name by all the industrial processes, all the transportation use, all the heating use, and so on. About 10,000 watts, 10 kilowatts.

Your own body produces only about 100 watts, 10th of a kilowatt. I'm going to use that round number off, 10-kilowatts, even though the number is actually almost closer to 12 kilowatts these days. About 10 kilowatts, 10,000 watts. You can only produce about one 10th of 1000 watts, one 10th of a kilowatt, 100 watts. Therefore, that 10,000 watts, 10 times would get you a kilowatt, so 100 times gets you 10 kilowatts.

The amount of energy used in your name is 100 times what your body can produce, and what that means is that if you had servants, turning cranks, there would be 100 servants working around the clock at devices like this, separate human servants, to produce the energy that you, one individual human being, use in 21st-century America.

Now, you might say, "Oh, but I drive a hybrid car, live in a solar-heated house, etc." Your number may be lower. For residents of my state, Vermont, surprisingly—because we are in such cold state—the number is more like 60 energy servants. The average though in United States, just averaged over all of us here, nearly 300 million of us, in total energy, our country consumes what comes out to about 10 kilowatts, or 100 energy servants' worth.

Let's take a look at some of them. Here you are, and here are some things that happen in your name, some energy uses. You drive a car. Maybe it's a big SUV, like that. You run light bulbs. You run a refrigerator, which we know requires energy, you run stoves, you cook, you probably have a big-screen television. You fly around in airplanes. There are factories working to produce goods for you. There are power plants generating the electricity you use. Even agriculture is not simple gardening, but a mechanized process that uses a large amount of energy. Far more energy goes into growing the crops than the food energy you and your body get out of them.

Those are your energy uses; and symbolically, that's the number of energy servants. There are the hundred energy servants it would take, turning those cranks, or doing deep knee bends, and converting that to useful work for you, to keep you going in early 21st-century industrialized society in the United States.

What do these servants do for us? What do they get us? Well, they do a number of things, and I'm going to break this down very crudely. These are statistics on what the energy servants do, how the energy is used. About 21% of our energy use is in the residential section. Heating and cooling our houses is the biggest single use. The next biggest single use is hot water heating, and it goes on from there. Refrigerators are a major user, although they're getting more efficient. Laundry machines, light bulbs, computers that we now have on all the time, televisions, that kind of thing. Twenty-one percent is residential.

By the way, since there are just about 100 energy servants, 21% means 21 of them are in your house, working for you. If you go home to your house, and there are just a few people in your family—but there are really, for everyone in your family—if you have four people in your family, there are 21 times 4, which is 84 servants around there. You better make room for them. Maybe three times as many, because the only work eight-hour shifts. You've therefore got about 250 people in your house, and you have to feed, clothe, and house them. They had better not use any energy, because you are going to need more.

About 21 of those servants are residential. About 18 of them are commercial. They're running the lights in the department stores, the refrigerators in the grocery stores, the checkout counter lasers in the scanners; they're running everything in the commercial sector.

The industrial sector uses about 33 of those servants. You say, "Well, I don't work in energy. I'm a teacher," but the industrial society is working for you, producing the goods that you use, and so, you are responsible for 33 servants working in industry, and about 28 of them in the United States are involved in transportation, most of them cars, some of them trucks, a smaller number of trains, and some airplanes. Most of that is from cars and trucks. That's what the servants are doing for us.

What kind of servants are they? Well, I drew them as little human beings, but of course, they aren't human beings, or we would be using the rest of the human race to satisfy our energy needs. Where do they come from? What kind of servants are they? What are the sources of our energy?

About 23 of those servants are coal servants. About 40 of them are oil servants. About 24% are gas servants. Add those up. You have 24 and 23 are 47 and 40 is 87%. Eighty-seven percent are fossil fuels. Those are the substances buried deep in the ground that consist of carbon fuels, fixed by sunlight in ancient plants, and that we are now digging out of the ground and consuming at a far greater rate than they were produced at, and dumping waste products—the primary one of which is carbon dioxide—into the atmosphere. That is causing some of the problems I alluded to in the second lecture of this module, "In the Greenhouse."

Where the rest of the servants come from? In the United States—this varies from country to country, and it varies dramatically from region to region. Eighty percent of them are nuclear where I come from, Vermont. We are the most nuclear-dependent state in the country, and all of New England is very heavily nuclear oriented. We don't have many fossil fuel powered plants, particularly in New England, but we do have a lot of nuclear power. If you go to the Pacific Northwest, there's a lot of hydroelectric power. That supplies only 3% of our energy nationwide, but it supplies a substantial amount in that area.

The rest are a remarkable amount—and this is the only number that has really changed significantly in the past few decades—wood and waste are now contributing about 4% of our energy, the "garbage to energy" plants. Then, there's "other." If you are a fan of solar energy, wind energy, geothermal energy, all of those good things, you have to look realistically here. They supply less than 1%, about 0.5%, of our energy. By the way, nearly all of that is geothermal, in a few plants in California, and that has its own problems. I'm not going to get into those in this lecture.

The fact is that the vast majority of our energy comes from fossil fuels, and that's true worldwide; something between 80 and 90% of the world's total energy comes from fossil fuels. As the developing world industrializes, more and more fossil fuels are going to be burned.

How do we get useful energy out of these fuels? Well, what we do with fuels is burn them. Well, I should back up. I showed you a fuel cell early on. Fuel cells aren't in widespread use yet, but if we had fuel cells going, there are ways to extract energy from fuels without actually burning them. Today though, we burn them. We produce heat, and we use heat engines.

I described in the last lecture, conceptually, what a heat engine did. Now, I'm going to describe a bit more practically how a heat engine works. Here's an example of what is a sort of typical heat engine. This is a particular kind of heat engine, called the *Carnot-cycle engine*. They're relatively few. There are no real engines that work exactly like this, but they work on the same general principle. What we have here is a cylinder. The gray thing is a movable piston. The yellow thing is a fluid that can expand or contract, and in some instances, maybe change state. In this one, though, it's just going to expand or contract.

What we do is take the cylinder, the piston, and the gas, and connect this to some kind of mechanical load. We connect the piston by means of some kind of shaft to a wheel that we want to turn that might be connected to an electric generator, or that might be the wheel of a train, or a car, or this looks a lot like a steam engine on an old steam train.

How do we extract energy? Well, we put the gas in contact with a hot substance, which we have heated up by burning one of these fuels, or fissioning uranium, or shining sunlight on it, or whatever. You can even think of a wind turbine as an example of this, when the heating is being done in the natural air circulation system.

The gas expands. As it does so, it pushes against the piston, and the piston turns the wheel, and does useful mechanical work. Then, we take the heat source away, but the wheel has some inertia, so it continues to turn, and the gas expands a little bit more. At this point, the wheel is still turning. It's got some inertia. It's going to turn just past that extreme motion of the piston. It's going to start to move the piston back. At that point, we put the cylinder in contact with the cool reservoir, because we have to cool that gas down. If we just let the piston turn, the gas will get compressed, and it will get hot. We've got to cool it back down, in order to let it expand again.

We put it in contact with the cool reservoir, and the gas compresses. That pulls the piston back, but as the gas compresses, and is in contact with the cool reservoir, it dumps some heat into that cool reservoir. That's the waste heat I talked about in the context of heat engines going to the environment. Then, we start the cycle over again.

That engine has extracted energy from the hot reservoir and has turned it into useful mechanical work, some of it; but in the process of getting it back to its cyclic place so that it can continue the action, it has had to dump some energy to the environment. By the way, the second law doesn't read exactly as I said in the previous lecture. It really says that it's impossible to construct an engine operating in a cycle so that it returns to the same state every so often, which is what all real engines have to do. It's impossible to build one of those with 100% efficiency.

You can work out—and an introductory physics student can do this—the efficiency of this engine. It's the useful energy you get out, divided by the total amount of energy you get from the heat source. So, if half the energy you extract from the heat source—from the burning coal, the burning oil, the burning gasoline, the fissioning uranium—if half that energy goes into useful mechanical work or electricity, and the other half get dumped to the environment as waste heat, you've got a 50% efficient engine.

It turns out that the engine's efficiency depends on the difference between the hottest temperature you have in the system—T-hot—and the coldest temperature you have, divided by T-hot. Now, that sounds a little bit quantitative, but it's not bad, and what it basically says is, the hotter you can make the temperature reservoir, the more efficient your engine is. You can't much change the low-temperature reservoir, because here on Earth, we are stuck with the ambient temperature, which is about 300 Kelvin. By the way, that formula works only for temperatures in Kelvin.

That is the efficiency of this engine called a "Carnot-cycle," and it's approximately the maximum efficiency of other real engines. The hottest temperature, unfortunately, is determined partly by engineering. You can't build a nuclear reactor without having to build a great big huge steel vessel to contain the reaction. The thickness of the steel we can easily work with. It determines how much pressure we can build up, and how hot we can make that, so practical engineering terms dictate how hot we can make that hottest temperature. They set the limits on the efficiencies of engines.

All real engines dump waste heat to the environment, and here's a picture showing two examples. On the left, there's a car radiator. You might think the radiator's job is just to keep the engine cool, because the engine happens to get hot. Well, the engine doesn't just happen to get hot. The second law says that it has to get hot, and waste heat has to dump to the environment. The radiator is one of the places it goes. It also goes out the exhaust pipe.

On the right, you see these typical cooling towers associated with power plants. Most people think they're just for nuclear power plants. That's nonsense. Any kind of big power plant has to have cooling towers like this. These cooling towers dump some of the waste heat to the atmosphere, avoiding dumping it to the rivers, and therefore, avoiding what's called *thermal pollution.*

Now, you might say, "Well, couldn't some clever engineer design a better engine, one that's more efficient?" The answer is, "no." That was proven in 1824 by a young French physicist named Sadi Carnot, and he proved that his engine, the Carnot engine that I just described, has the maximum possibly efficiency, and that no engine that you could build could have more efficiency than that, and you might build an engine with less efficiency. Thus, that Carnot engine, with that efficiency—it depends on the maximum temperature and the minimum temperature—that's the maximum possible temperature that you can get; and as a result, our typical power plants, for example, range from somewhere around 25% efficient, to roughly 50% efficient for the most advanced cycle power plants we have today. More than half of the energy we make in all of our power plants is dumped as waste heat.

By the way, most of the flow of freshwater that lands as rain on the United States, and flows to the rivers—most of it, sooner or later, finds its way through the cooling systems of electric power plants. It's that big a deal, and we sometimes cool ahead of time with these cooling towers, and other things, to keep that waste heat out of the environment, or it would be thermal pollution.

Couldn't we do something with that waste heat? The answer is that we could, and again, from the previous lecture I mentioned that the second law is really about energy quality, rather than energy quantity. Some of our energy uses require relatively low quality energy. Heating water on the stove. Heating water to take showers. Other forms of energy, like transportation, electric motors to run our disk drives in our computers, and our CD drives, our electric clocks, and our washing machines—those require the highest quality of energy.

The second law of thermodynamics says that we can't make that high-quality energy from thermal sources, from burning fuels with 100% efficiency. Here's the deal, though. In the United States, we tend to use energy—roughly one third of it—as this high-quality energy motion electricity. We have to have that. About a third of it is high temperature heat for industrial processes—steam, usually—at greater than 100 degrees Celsius. About one third of it is heat for more mundane purposes, like cooking, heating houses, and so on—heating buildings, at less than 100 degrees Celsius.

The energy we produce, on the other hand, with our power plants, particularly, and our cars, is typically high-quality energy, and less of it is heat. There is therefore a kind of mismatch there. We're producing more high-quality energy than we need, and sometimes, we do things that are really not very energy efficient.

For example, let's look at the example of heating hot water. Here's an electric hot water heater. Now, we're going to add one energy unit to this hot water heater, whatever it is—a joule, a kilowatt-hour, or whatever unit of energy we are going to add to this hot water heater. Now, electrical energy is very efficient in the sense that if I stick an electrical resistance heater in a water heater, virtually all of the energy I send into that electrical resistance as electricity comes out as heat in the water. It's almost 100%. Thus, if I supply one unit of electrical energy to that water heater, I get one unit of heat. It sounds great.

However, a power plant, which is typically on the order of 33% efficient, a little bit more for advanced power plants, and a little bit less for some of the older ones—nuclear plants are a hair less efficient, not because they are nuclear per se, but because it is, again, harder to achieve the very high temperatures when you try to contain the nuclear reaction.

Typically, then, something like three units of fuel energy—coal, nuclear fission, uranium, or whatever—would be used to produce that one unit of electricity. Therefore, this water heater, although it looks great, is really only 33% efficient. It takes three units of fuel to produce one unit of heat.

On the other hand, here's a gas water heater. There's a flame right in the gas water heater. The gas water heater is very efficient. A modern gas water heater is 80 to 90% or more efficient. What that means that it takes perhaps 1.05 energy units of gas— stored energy in the gas fuel—to produce one unit of hot water. That is a much more efficient use of our energy. We should try to match the end uses of our energy to the quality of the energy generation source. If we did just that, we would have all the energy we still use. We would not reduce that at all, but we would use that energy in more efficient ways.

By the way, some industries in the United States are increasingly doing this. They are developing what is called *cogeneration*"—I hinted at this briefly in the previous lecture. Instead of buying electrical energy from the power companies and then having a boiler to make the steam you need for your industrial processes, you take a boiler, make the steam, and then use the steam to turn your own electric generators, extracting some of the energy as electricity. You use that for your high-quality needs, and what would have been the waste heat from the electric power generation becomes the lower quality energy you need for your steam, or heating your buildings, or whatever. More and more people are cogenerating. Currently, Europe, which is the world leader, is approaching 30% electricity by cogeneration, so that they are much more efficient in matching their end uses to the energy quality.

Another method of using energy more efficiently, as I mentioned briefly in the previous lecture, is to use heat pumps. The heat pump is basically a refrigerator in reverse. We might extract, say, five units of heat from the ground, dump that out of the heat pump into the house, we bring in one unit of electrical energy to do it, and all of that ends up as heat. One unit of electrical energy has helped move another five units of heat from the ground, giving us six units of heat in the house. That is a heat pump, which is another way of handling energy efficiently.

Applied to these heat pumps, the second law basically says, "Yeah, you've got to supply some extra energy, some useful mechanical work, but you do so in a way that transfers much more energy than the energy you are bringing in as electricity."

Let me end by saying that the examples in this lecture so far represent ways to use the energy we use more efficiently, to get those 100 energy servants from less fuel. However, do we really need all of those 100 energy servants? Do we really need to drive cars that go 10 miles on a gallon of gas, because they feel big and militaristic to us? Do we really need to take our clothes out of the washing machine and throw them in the dryer, when just a step outside, hanging them in the breeze, can dry them with no use of energy?

By the way, I've calculated in a little book I wrote on nuclear issues that suppose every household is using a clothes dryer one hour per day. If we decided to hang our clothes up instead, we could produce 200 nuclear bombs' less worth of plutonium in the nuclear power plants of this country every year, just by that step. The ways in which we choose to use energy have consequences, and I think that personally, we could not only use our energy more efficiently, but we could reduce the number of these energy servants we need.

Here's my picture again of you, surrounded by some of the energy users that you use, and those 100 energy servants that are cranking away for you to produce that energy. What could you do? Well, you could replace that SUV with a hybrid car. You could replace some of that electrical energy generation by wind. You might replace your big-screen television set—I guess I didn't do that here—with an LCD screen that uses less energy. You might ride the train instead of an airplane. Trains are much more efficient.

This factory has fewer smokestacks. It's running more efficient processes, polluting less in the process. You've got an energy efficient light bulb. You are going out and growing some of your own vegetables with your gardening tools, and if you did all those things—we don't all have to go back to the land and become hippies, and grow organic foods, but if we did just some of these things, like buying different cars, or buying green energy when it's available to us—if our electric company says, "Gee, if you pay 20% more, we guarantee that your energy is coming from renewable sources," would you do that? It would sure help. If you did that, I think you could get rid of, let go, at least half of those 100 energy servants. That is energy in your life.

Module Six: Potpourri
Lecture Thirty-One
Your Place on Earth

Scope: This module begins with "Your Place on Earth" and ends with "Your Place in the Universe." That final lecture is a grandiose look at your intimate physical connection to the broader Universe and its evolution. This first lecture is, in contrast, technologically specific. Here *place* means *position*—specifically, how you can know your exact position on Earth. Space technology, in the form of the Global Positioning System (GPS), provides the answer. GPS is an increasingly pervasive and, in some cases, invasive technology. Originally developed for military purposes, GPS is now used routinely for sea, air, and even automobile navigation; for tracking vehicles, freight, animals, and people; and increasingly, for emergency location of 911 calls from cell phones. GPS also aids surveyors, geologists, environmental scientists, oceanographers, and others in studying our planet. The system represents a remarkable confluence of physics-based technologies, from orbiting satellites to the exquisite timekeeping of atomic clocks. In the coming years, GPS will almost certainly play an increasingly important role in your life.

Outline

I. In space, GPS consists of a "constellation" of nominally 24 satellites, in orbits about 12,000 miles above Earth, where the orbital period is 12 hours. At this altitude, there is essentially no air resistance; thus, the satellites' orbits and, hence, positions are known with exquisite accuracy.

   A. The relative positioning of the orbits means that every point on Earth can always see at least four satellites.

   B. The satellites' exact orbits are monitored from the ground, and updates are frequently sent to the satellites. The satellites continually broadcast their precise positions, as well as the time.

C. GPS satellites have lifetimes of about 7–15 years, and older satellites are regularly replaced by new ones with enhanced capabilities.

II. GPS works by measuring the distances from satellites to a given GPS receiver. You may once have learned to determine distances by triangulation from a known position; GPS is somewhat the opposite, getting position from distances.

    A. Knowing the distance to one satellite tells us that we're somewhere on a sphere whose radius is that distance (in two dimensions, the sphere would be a circle).

    B. Knowing the distance to two satellites tells us that we're on the intersection of two spheres—namely, somewhere on a circle (on one of two points in two dimensions).

    C. Adding the distance to a third satellite narrows the choice of position to just two points—one of which can usually be rejected because it makes no sense.

III. Distances are determined by accurately measuring the time it takes radio signals, moving at the speed of light, to travel from satellite to receiver.

    A. If both satellite and receiver had perfectly accurate clocks, signals from three satellites would suffice to determine position.

    B. The satellites have extremely accurate (and expensive!) atomic clocks on board.

        1. Atomic clocks are based on the frequency of light emitted as electrons jump from one atomic energy level to another. One second is defined as exactly 9,192,631,770 cycles of the light emitted or absorbed in a particular electron transition in the element cesium. The atomic energy levels used are very close, corresponding to so-called *hyperfine transitions*, and the associated frequencies are in the microwave region of the electromagnetic spectrum. Atomic clocks use cesium or other elements to implement an atom-based time standard.

        2. The clocks used in GPS satellites are typically accurate to about 10 billionths of a second, and each satellite carries four clocks. Because of its satellite-based atomic clocks, GPS can be used to determine time as well as position.

3. Each satellite sends out a unique signal consisting of a seemingly random ("pseudo-random") string of digital 1s and 0s. The receiver generates its own versions of these signals, which should be in sync with the satellites's, but because of the distances to the satellites, there are delays. Measuring those delays determines the distances.

C. GPS receivers would be prohibitively large and expensive if they contained atomic clocks; for this reason, they contain less accurate clocks.

   1. As a result, position measurements based on three satellites are not accurate.
   2. GPS receivers use measurements from a fourth satellite to correct for receiver clock errors. Because three perfect measurements alone would fix the receiver's position, any inconsistency with a fourth measurement contains information about the receiver's clock errors; with the fourth measurement, then, those errors can be corrected.
   3. The result is a measurement of position on Earth that's typically good to tens of feet and can, under some circumstances, be accurate to mere inches.

D. In addition to position, GPS receivers also determine relative velocity between the receiver and satellite using the Doppler effect (Lecture Six). This permits faster and more accurate calculation of the receiver's motion.

IV. The GPS system corrects for a number of other errors, some trivial and others profound.

A. The satellites's locations are known but not perfectly. Position predictions are updated approximately each hour and are typically good to a few meters, or about 10 feet.

B. Because the satellite clock times are typically good to about 10 billionths of a second and because light travels about 1 foot in a billionth of a second, clock errors introduce a distance error comparable to the 10-foot error in satellite position.

C. Earth's upper atmosphere, or *ionosphere*, contains electrically charged particles (in particular, electrons) in varying abundances. These alter the speed of the radio signals from GPS satellites and, thus, the calculated satellite distances. Although knowledge of atmospheric conditions can help correct these errors, the most sophisticated correction comes from the fact that different radio frequencies are affected in different ways. GPS satellites broadcast on two separate frequencies, and by comparing the arrival times of the two frequencies, a receiver can reduce this error to approximately 15 feet.

D. GPS signals may reach the receiver directly or after reflection from water, buildings, or other objects (this same phenomenon causes "ghost" images in TV pictures). Calculating distance based on a reflected signal would lead to error. Furthermore, these "multipath" signals interfere with each other. GPS receivers use sophisticated schemes for determining which signal arrives first, in an attempt to use only information from the direct signal. Multipath error typically amounts to about 2 feet.

E. A particularly fundamental source of potential error arises because of Einstein's special and general theories of relativity. According to the special theory, the relative motion of satellites and observers means their clocks are not measuring the same time. The general theory deals with gravity, and in the GPS context, it says that time at the satellites's altitude runs faster than on the ground. Both these effects must be taken into account, or GPS measurements would be off, in a day's time, by close to a mile.

F. *Differential GPS* is used in precision applications to reduce errors still further. Differential GPS uses two GPS receivers that may be up to a few hundred miles apart.

1. One receiver is placed at a fixed point whose location is accurately known. The second roving receiver is then used to make measurements relative to the first.

2. Because the receivers are fairly close, signals reach them through the same atmospheric path; therefore, atmospheric errors are the same. Given that one receiver is in a precisely known location, the discrepancy in timing due to the atmosphere can be determined and accounted for in the roving receiver.

3. Similarly, the fixed receiver can determine the clock and position errors for the individual satellites and relay this information to the roving receiver.

4. The use of differential GPS essentially eliminates clock and position errors and reduces other errors to a few feet. More sophisticated differential GPS techniques can push errors down to a few inches.

V. GPS is widely used in a variety of applications; here, we sample just a few.

 A. GPS reception is increasingly built into cell phones to provide 911 emergency location information, as well as commercial enhancements, such as providing phone numbers of nearby restaurants, as determined by the phone's GPS location.

 B. Hand-held GPS receivers provide position information for hikers, boaters, and others. Coupled with built-in map software, these units display the user's position as a location on a map.

 C. GPS navigation systems in cars use GPS to monitor continuously the car's location. The information is combined with optimizing software, digital road maps, and computer voice technologies to determine the best route to a desired destination, then to provide real-time voice and/or visual instructions at each turn. In tunnels or when tall buildings obscure the GPS satellites, a less accurate system based on the vehicle's odometer and acceleration sensors temporarily keeps track of position.

 D. In GPS vehicle-tracking systems, a GPS receiver on each vehicle relays the vehicle's position to a central monitoring station. Uses include tracking freight-carrying trucks, deploying emergency vehicles and taxis efficiently, and even determining if rental cars stray beyond their allowed boundaries. Anxious parents or covert operatives can install GPS receiver/recorder units surreptitiously, later downloading detailed logs of a vehicle's travel, including street names!

 E. In farming, GPS guides tractors to accuracies of an inch or so, reducing fuel consumption and time involved in plowing. Uses like this require differential GPS, but because only relative positions are important, the fixed GPS need not be at an accurately known location. A similar technique is used to guide heavy construction equipment.

 F. Because GPS gives altitude as well as longitude and latitude, it's ideally suited for aircraft navigation. Use of GPS can help alleviate air traffic congestion, allow more efficient flight paths, and augmented by ground-based systems, bring all-weather capability to smaller airports.

G. GPS has many uses in science and engineering. Here are a few:
  1. Using differential GPS, geologists monitor the flow of glaciers and even the growth of mountains.
  2. Oceanographers measure the scattering of GPS signals off the ocean surface and use the result to calculate wind speeds at the ocean surface.
  3. Engineers monitor deformation in such structures as dams and bridges, again, using differential GPS to provide inch-level accuracy.
  4. Wildlife biologists place GPS collars on large mammals to monitor ranging behavior and herd migration. The devices can log months worth of data for later analysis.

H. Although you may not realize it, GPS is probably already an example of physics in your everyday life. And GPS will only become more prevalent in the future.

Suggested Reading:

Marshall Brain, *Marshall Brain's How Stuff Works*, pp. 91–93 and 95.

Per Enge, "Retooling the Global Positioning System," *Scientific American*, May 2004, pp. 91–97.

Going Deeper:

Ahmed El-Rabbany, *Introduction to GPS: The Global Positioning System*.

Bruce Grubbs, *Using GPS: GPS Simplified for Outdoor Adventurers*.

Questions to Consider:

1. Measurement of the distances to three satellites should be enough to fix one's position. Why, then, must GPS receivers "see" a minimum of four satellites?

2. Should accurate GPS signals be provided, as they are today, essentially freely to anyone on Earth?

Lecture Thirty-One—Transcript
Your Place on Earth

Welcome to Lecture Thirty-One, "Your Place on Earth." This is the first lecture in the final module, Module Six, entitled "Potpourri," and it is a real Potpourri of different ideas, technologies, physics concepts, and physics applications. Some of the lectures in this module are outgrowths of some of the earlier modules. They are, perhaps, more complex material that wouldn't fit into those other modules. Most of them draw on ideas from several of the earlier modules and put together concepts that we wouldn't be able to deal with, with the subject matter in one of those modules alone.

Module Six is bracketed by two lectures whose titles sound similar. They both begin with, "Your Place." You therefore might think that they're kind of similar. The first lecture, this one, is, "Your Place on Earth." The final lecture is, "Your Place in the Universe." They therefore sound kind of similar.

Nothing could be more different. This lecture, Lecture Thirty-One, is about a specific technology. It's probably the most specifically technological, and, in a sense, the most contemporary, technology I'll be dealing with in the entire course. It's a very specific lecture about a very specific technology that allows us to determine our place on Earth.

Lecture Thirty-Six, the final lecture, "Your Place in the Universe," is a much more grandiose, philosophical look at where we human beings fit into the Universe, in particular, how the material of which we are made is descended, ultimately, from the stars, and from the processes that began in shape our Universe.

In between, we have more of that hodgepodge-like potpourri. Lecture Thirty-Two, "Dance and Spin," deals with rotational motion, something that might have fit into the second module, but didn't fit, and that also has some specific implications ranging as far and wide as how the North Star will not always be the North Star, to how medical resonance imaging works, the fundamental principle behind it.

Lecture Thirty-Three, "The Light Fantastic," talks about lasers. I said at one point that the transistor may have been the most important invention of the 20th century, but the laser ranks right up there as well, and Lecture Thirty-Three deals with lasers, how they work, and we will see a number of demonstrations of lasers in action.

Lecture Thirty-Four is, "Nuclear Matters," a little bit about the atomic nucleus, its role in producing energy, its role in a lot of other important technologies, some of the dangers it poses, and so on.

Lecture Thirty-Five is "Physics in Your Body." We will start out with a little bit of the mechanics of how your body works, but most of that lecture will be dedicated to medical techniques that use physics, particularly, medical imaging: PET scans, CAT scans, MRI scans bone scans, and that sort of thing. It therefore is a real potpourri of different lectures. We will be shifting gears fairly rapidly, from one to the other.

Lecture Thirty-One, this one, "Your Place on Earth," is about the global positioning system. The global positioning system is quite a new technology. It's been available civilians in its full form only since the year 2000, and it only got into full implementation in about the year 1994, so it is a quite new technology, and is already revolutionizing many, many areas of business, commerce, technology, science, and so on. We are therefore going to look, in this lecture, at the global positioning system.

In a sense, the global positioning system is an outgrowth of Module Two, when we had, again, the lecture on "Into Space," because the GPS system is based on satellites. Satellites or spacecraft, and they obey all of those orbital parameters that we talked about in that module. However, there are many other technologies involved in the global positioning system.

Let me begin with a brief history of the global positioning system. It's a brief history, because, as I indicated, this is a fairly new technology. The first trial satellites for GPS went up in 1974 and 1977. GPS was initially conceived as a military project, and until the year 2000, it was, in some sense, exclusively a military project, in the sense that civilians were not allowed access to the most precise measurements it could provide.

In 1978 to 1985, the so-called "Block I" satellites were launched. These were the first generation of GPS satellites. Satellites were being continually improved. From 1989 to 1994, the "Block II" satellites, some of which are still up there—more recent versions of them—were launched, and in 1994, the system became fully operational, meaning that the 24th satellite was launched. GPS requires that there be at least 24 satellites up there at all times.

In 2000, so-called *selective availability*, which is what made the civilian use limited in its accuracy, was abolished, and so, the full GPS accuracy became available for the first time to civilians.

What is this global positioning system? Well, it's primarily—not exclusively, but primarily—a space-based system of the so-called *constellation* of 24 satellites. This is a symbolic picture of what it may look like. I haven't shown all 24 satellites, but there is this swarm, called a "constellation," of satellites. There have to be at least 24 at all times, and there are usually a few spares up there, in case one of them fails.

They are up about 12,000 miles, in inclined orbits—I discussed the inclination of orbits back in the lecture on space flight—these orbits are inclined at 55 degrees. That's a fairly high inclination, and it allows them to be seen from anywhere on Earth. The orbital shape is circular, and the orbital period is about 12 hours, so these are going around about twice as fast as the Earth is rotating underneath them. Those are the parameters of the GPS constellation.

Now, the exact orbits are monitored from the ground, so we know exactly how the satellites are moving. Furthermore, because they're up at 12,000 miles, they have essentially no air resistance, and their orbits are very accurately predictable. We monitor these orbits, and occasionally—actually, more than occasionally; sometimes as often as every hour—we send information to the satellites, telling them exactly where they are, so satellites the know that information. Then, given that the orbits are very predictable by Newton's laws, they continue to know very accurately where they are for some time to come, and they continually broadcast signals that carry information about their precise positions.

GPS satellites typically last somewhere on the order of 7 to 15 years, and then they are continually being replaced. Then, newer satellites tend to have better capabilities, more accuracy, and so on.

How does GPS work? Fundamentally, it works by measuring the distances to, in principle, three, in practice, four—and I will tell you why it's three versus four—of the satellites, and uses the measured distances to determine exactly where on Earth an observer is.

What a GPS user sees is the following. The GPS user is sitting here on Earth, and the system is set up. The constellation of satellites is configured so that a given observer can always see at least four satellites. There are always at least four satellites in the sky above you, if you are using GPS, and the four satellites are always above the horizon. They're not always the same four satellites, but you will always be in view of four satellites. The question is, how far is it to each of those four satellites? If you can answer that question, as I will show you, you can pin down your position with incredible accuracy.

Let's see how that works. You have probably used at some point, or learned about, a technique called *triangulation*, to measure distances. For example, I was talking to a biologist who was measuring the heights of the trees in a tropical rain forest, and the way that was done was to measure the distance from the base of the tree to the observer, and then sight up to the top of the tree, and then, knowing that angle and a little trigonometry, you could calculate the height of the tree. That's triangulation.

GPS uses what is kind of the opposite. It finds out where you are by determining distances. Translation determines distances by knowing where you are, and some other measurements. The GPS satellite measurements determine where you are from distance measurements, and that's a process called *trilateration*.

I'm going to describe that process, and I'm going to describe it first in two dimensions, because it is easier to understand what's going on in two dimensions. I've got a two-dimensional screen, and so, I'm going to describe the picture in two dimensions. I will then show it to you in three dimensions, which gets more complex, and occasionally throughout this discussion, I will drop back to two dimensions just to make the picture simpler, but remember that all of this is going on in three dimensions.

How does this work? Well, here's a GPS satellite, and suppose I somehow know how far I am from that satellite. I will show you shortly how we know, in fact, how far we are from the satellite. Well, that tells me is the distance. I might be at the end of that line. I might be the end of that equal length blue line. I might be at the end of that equal length blue line. In fact, again, in two dimensions only, I might be anywhere on a circle whose radius is this known distance to that satellite. Thus, all that distance measurement tells me—if I know that I'm 12,340 miles from that satellite, all that tells me is that I'm somewhere on a circle, and in three dimensions, it would be a sphere whose radius is that distance. It doesn't tell me where on that circle, though.

Remember, though, that there are more satellites in my field of view. Here's another satellite. Maybe I'm a little farther from that satellite, so that I know I'm on a circle whose radius is the distance I know I am from that satellite, and in two dimensions, those two circles intersect only at two points. There they are, and so I know in principle that I made one of those two points. It turns out that the ambiguity between which of those two points I'm at can usually be solved. One of those two points will typically be up in space, off the surface of the Earth, or maybe will be down inside the Earth. One of them, though, is on the surface of the Earth, and that's enough to tell me where I am in two dimensions.

Trilateration, then, in two dimensions, requires, in principle, measurements of the distances to two different satellites, and that pins down my position to within two possible points, one of which can usually be rejected, because it's obviously absurd.

Well, the world is not two-dimensional. It's three-dimensional, so we need to do this process in three dimensions. That's a harder picture to draw, but let me give you a sense of how that works. Here we are, know, in three dimensions. Again, here's a GPS satellite. Here's the sphere surrounding that satellite, because in three dimensions, if I know the distance from myself to the satellite, all I know is that I'm on a sphere in three dimensions, on the surface of a sphere whose radius is the distance that I know to that satellite.

Well, here's a second satellite, the same one I was talking about before, but now, instead of being on a circle about that satellite, all I know is that I'm on a sphere whose radius is equal to the known distance of that satellite. Again, though, I don't know where I am on that sphere. Put together the information from those two satellites, and you can see from those two pictures that the two spheres intersect in a circle, so now what I know is that I'm on a single circle.

Three dimensions makes things more complicated. In two dimensions, two satellites were enough to tell where I was within this ambiguity. In three dimensions, all they do is put me on a circle, what one satellite did in two dimensions, but the world is three-dimensional, so here's where we are.

We need a third satellite. If there's a third satellite, and I know I'm a certain distance from that satellite, then I know I'm also on a sphere that surrounds that satellite, and whose radius is the known distance to it. If you look at the picture, you'll see that those three spheres ultimately intersect in two points. Then, with three satellites, I know I'm on one of those two points. Again, one of them is usually absurd—it's way up in space, it's down inside the Earth, or something—so that we can easily pick out which one I'm on.

That's great. That is, then, how, in principle, we determine our position on the surface of the Earth. How do we determine it to within a few feet? It's where we are going in the rest of this lecture. It's a little more subtle than just doing this, as we will see.

How do we determine the distance? Well, the distances are determined, ultimately, by timing radio signals coming from the satellites, and using the known satellite to calculate the distances to the satellite. The speed of light is fast, but we can nevertheless use it for timing. If we had perfectly accurate clocks in the satellite and on the ground, in the receiver we use, we could, in fact, measure that time with exact precision, and with the exception of some problems I will bring up in a minute, we can determine our position.

Now, the satellites have on board very, very accurate comic clocks. They're good to about 10 billionths of one second. I want to digress from moment about how atomic clocks work, because they are the essence of the GPS technology, in a sense. They're also the essence of how we keep time here on Earth, in general.

Let me just take a little digression and tell you what an atomic clock is. And atomic clock is a rather sophisticated thing. It consists, typically, of a generator of microwaves, some kind of cavity into which those microwaves are beamed, non-uniform magnetic field—I will tell you in a minute what that does—and an atom detector.

Now, our standard of time is, in fact, based on transitions among energy levels in atoms. I talked about such transitions back when I was talking about how semiconductors work, for example. In particular, the atom cesium undergoes a so-called *hyperfine* transition between two very closely spaced energy levels; and in the process, it emits electromagnetic radiation at a precise frequency. Our definition of the second, what one second means to humankind these days, is 9,192,631,770 oscillation cycles of the energy associated with that transition in the atom cesium. That's how a second is defined, and atomic clocks reproduce that transition.

Here's how it works. We send a beam of atoms through this cavity, and if the atoms get bumped up into that slightly higher energy level, which they will do if we bombard them with microwaves of just the right frequency, then, when they pass through that non-uniform magnetic field, they will be deflected upward into that atom detector. If they haven't been bumped up into that slightly higher energy level by microwaves of just the right frequency, they will be deflected downward, so we do is have a control, a feedback loop, that looks at the number of atoms coming into that detector, and changes the frequency of the microwave generator, which is what is bumping those atoms up, in such a way as to maximize the number of atoms coming in to that atom detector.

When we get the maximum number of atoms, we know we've got the microwave frequency at exactly the atom frequency the atoms want to undergo that transition at. Now, there are some subtleties. The atoms have to be cold, because we actually dropped them down into the tower, but ultimately, we connect a digital counter to that microwave generator. We know those microwaves are at the right frequency to excite that transition in the atoms, and so, we are there, right at that frequency, and from that, we can actually tell the time, the United States standard time, which you can get off your computer.

You can, in fact, synchronize your computer's clock to the standard, or maybe you have a so-called *atomic clock* at home. It's not an atomic clock, but is simply a clock with a receiver. It gets the signal from the US standard atomic clock.

You are looking at signals from this clock. This is the US official time standard: the NIST F-1 cesium atomic clock. "NIST" is the National Institute of Standard and Technology. Here's their F-1 cesium atomic clock. This is a bigger, more sophisticated clock than those on the GPS receivers, and it is what sets the standard of time for the United States.

Now we know how atomic clocks work—back to GPS. GPS satellites carry four different atomic clocks. Some use cesium, some use rubidium, but they basically work on the same principle. By the way, GPS is then broadcasting a time signal, so it can also be used to set your time, and GPS receivers know what time it is, because of that.

Each satellite sends out a unique signal that identifies that satellite. It's just so-called *pseudorandom code*, a string of ones and zeros. Here's what happens when you are using GPS to try to figure out where you are. There's a particular satellite up in space. It's over the horizon, and you can see it. The receiver knows these pseudorandom codes of all the different satellites, so you receiver identifies what satellite that is, and it produces in itself this pseudorandom code.

You get the pseudorandom code from the satellite, but it comes to you a little bit delayed, and that time delay tells you how long it took the signal to get from the satellite to you. That's what determines your distance from that satellite. Remember that we had the known distances to three satellites.

Here's a little math. The speed of light is about 186,000 miles per second. The satellites are up about 12,000 miles. Therefore, the time it takes to get from the satellite to you is about 12,000 miles, divided by 186,000 miles per second. That's about six to seven hundredths of a second, about 65 milliseconds, a pretty small time.

In those distances, if that timing were off even by 1000th of a second, at 186,000 miles per second, your distance would be off by 186 miles. The GPS is good to 10 feet, or sometimes to inches. How do we achieve that?

Well, we've got to look into a few other things. First of all, your receiver is not a great big heavy thing with an atomic clock in it. We can't afford to put an atomic clock in, or afford to fit one in receiver, so it has a more standard quartz clock, like your quartz watch. We have to now correct for the fact that that clock is not really good enough. Here's how we do it—and this is a two-dimensional analogy. Remember that if we had two satellites, and distances to them, we would know exactly where we were in two dimensions, or we would know to within those two points.

What if we had other satellite as well? Well, presumably, its circle would also have to intersect at that point, because we are at a specific place, after all. This is therefore the ideal. This is a perfect agreement, but because the clock in the receiver is not that accurate, there's some error. There is some uncertainty in the distances, because there's a clock error in your receiver.

What do we do? Your receiver cleverly looks at the error; only it does this with four satellites, because it took three satellites in three dimensions to pin us down. With the clock error, that's not good enough. We take the measurements from a fourth satellite. If they are inconsistent, which they will be because of these timing errors, we then recalculate. Your receiver then recalculates all of the distances, in order to find the unique solution that puts us at the same place, and then, your clock is corrected, and matches the GPS transmitter clock in the satellite. That's why we need four satellites—three in two dimensions, four in three dimensions—and the result is a measurement that is typically good to within a few tens of feet, and I will get to that in just a moment.

In addition to position, GPS receivers also measure the relative velocity. They do that by the Doppler shift, so that if you are driving along in the countryside, the GPS receiver can calculate how fast you are moving.

There are a number of other errors that we have to correct for, or at least take into account. For instance, the satellites' positions are known, but not absolutely perfectly. They're typically good to about 10 feet, so that there's a 10-foot error. The clocks in the satellites are good to about 10 billionths of a second. Light travels about one foot in a billionth of a second, so there's another 10 feet.

A more subtle error is Earth's upper atmosphere. The satellites are up at 12,000 miles. We know the exact speed of light, 299,792,458 meters per second, exactly by definition. We know the speed above the atmosphere. In the atmosphere—only about 100 miles or so, depending on what angle we are coming at—the speed is affected by atmospheric conditions. Here, I want to draw on the module on electromagnetic magnetism a little bit, because here's why this happens.

About 50 miles up, there's a layer of atmosphere where high-energy radiation from the Sun—ultraviolet and so on—have ionized some of the atoms, and there are a number of free electrons around there. The Earth has a magnetic field, and those electrons are doing what electrons do in a magnetic field. They are experiencing magnetic forces. They're charged particles, moving charged particles, and they go spiraling around the magnetic field lines. That has a profound effect on radio waves coming through. Those rotating electrons interact with radio waves and slow them down. The amount by which they slow them down depends on how many of these electrons there are, and that depends on local conditions, what the Sun is doing, and so on, and is quite variable.

That's a hard thing to compensate for. However, fortunately, that effect is frequency-dependent, and satellites broadcast signals at two different frequencies, so here comes signals into different frequencies. They come out, out of step with each other. Each one has been slowed down by a different amount, and by determining that amount, we can determine what the ionospheric conditions are, and we can sort of correct for that. Our corrections bring that error to about 15 feet.

Another thing that happens is that signal from the satellite is supposed to come directly to your receiver, but what if there's a tall building nearby, or a body of water, or something? The signal bounces off and comes to receiver. This is sort of like when you see ghost images on TV. The ghostly image you see, maybe to the right, is from a weaker signal that has come a little bit later into your TV receiver by bouncing off some object in your environment. This multipath error is a problem, and it typically amounts to about two feet. The satellite receiver has sophisticated algorithms for trying to figure out which signal got to it first.

There are a variety of error sources in GPS, therefore: The position is about 10 feet; the satellite positions, clocks, are about another 10 feet; atmospheric error, about 15 feet; multipath error, about another two feet. The result is that typical GPS accuracy is a few tens of feet. You know where you are on the surface of the Earth to a few tens of feet.

By the way, there's another subtle effect, or sophisticated effect that would cause errors, except that we know to correct for it. Both Einstein's special theory of relativity and general theory of relativity talk about things that happen to time when an object is moving relative to you. The satellites are moving relative to you, not fast compared to the speed of light, but still, pretty fast. They're up high, where gravity is different, and both special and general relativistic effects have to be taken into account. If they weren't, by the end of a day, the measurements made by GPS satellites would be off by about a mile. There's one case where Einstein's theories of relativity are really everyday physics.

I've got us down to a few feet, but we can do much better than that, and we do that with so-called *differential GPS*. Here's how differential GPS works. You have a second receiver. You put it at a fixed location, and a known location. The location has been surveyed, so we know exactly where on Earth this location is, which may be up to 100 miles or so from where you are doing your work. It's a known location, and is close enough to you that atmospheric conditions above you and that receiver are the same, roughly, closely. We can therefore compensate for that error completely. We get rid of that atmospheric error completely.

Furthermore, because this receiver knows exactly where it is, the errors in its measurement, as determined by GPS, can correct for those errors, and it relates all those corrections to you in your nearby roving GPS; and with differential GPS, you can easily get down to a few feet. With really sophisticated techniques, you can get down to a few inches. That's differential GPS.

What do we use this GPS for? I'm just going to give you a brief introduction to the many, many, many ways GPS is used in today's world. For example, GPS is increasingly built into cell phones, so that if a cell phone call is made to 911, an emergency call, the 911 operator immediately knows where the cell phone is. That was a weak link in the whole 911 system. When we got landline phones, so that the 911 operators knew where they were, that didn't work for cell phones. Well, cell phones can identify themselves, possibly by triangulation and a cell phone towers nearby, but with GPS, anywhere on Earth, a cell phone can figure exactly where it is, and send that information to 911.

Handheld GPS is great for hikers, boaters, and so on. Here's a picture of a hiker with a handheld GPS receiver, looking off in the mountains, and knowing exactly where he is. We use GPS navigation in cars. Increasingly, cars are coming with built-in GPS navigation systems, or you can buy them as add-ons. They include things like that that show exactly where you are, and some of them will even speak corrections to tell you where to turn. If you go into a canyon, between buildings, and through a city or tunnel, there are auxiliary systems based on your car's odometer and acceleration measurements that continue to tell you where you are until you are back in view of the satellites again.

We use GPS for tracking for a variety of applications. For example, a freight company might want to know exactly where its trucks are at all times, so its trucks are equipped with GPS receivers. The GPS receiver knows where the truck is, and then, transmitters on the truck beam it to the company's headquarters. It knows exactly where its trucks are and when they're going to arrive at their destinations, if its drivers are goofing off by the roadside rest area destinations, or whatever.

There are a whole other bunch of applications for vehicle tracking. Emergency vehicles and taxi fleets are tracked. That provides the most efficient deployment of vehicles to where they are needed; for example, rental car companies are increasingly using GPS. I recently read a case where a person had rented a car in California. The rental agreement said that you had to stay within the State of California. The renter decided that they wanted to go to Las Vegas, which is not very far from the California border, but far enough so that the GPS device in the car recorded that stray over the state line, and they were hit up with a fine or something, when they got back.

If you are an anxious parent, you can actually have a covert GPS recorder installed in your car, if you want to know where your teenagers are going with the car, and it will record every movement of that car. It can come with maps of your area, so that it will tell you, when you play it back, exactly what street they've been on. You download this thing into your personal computer. If you're some kind of covert operative who wants to track somebody, these things are being done now.

Farmers like GPS. Farmers waste a lot of fuel plowing over the same ground twice, or overlapping their plows, or whatever. Farmers can use GPS. They reduce fuel consumption when they do that, and they also can reduce the time it takes to plow a field. Here's a picture showing, symbolically, a tractor pulling a plow, plowing these nice straight furrows.

How does that work? Straight GPS tells the tractor where is to within a matter of a few tens of feet. That is certainly not good enough to plow a nice, straight row, get the next one parallel to it and not overlapping, but not leaving any gap in between, or getting it positioned where you want it to be, so you set up a differential GPS.

In this case, you don't need to know exactly where the differential GPS transmitter is. All you've got to know is where you are relative to it, and that's really easy. Here's a differential GPS receiver at the edge of the field. It gets that information and then it becomes very easy to measure the tractor's position to within inches relative to that differential GPS receiver. That allows the farmer, really, to save a lot of fuel—cutting down on those energy servants that I talked about in a previous lecture, by the way—if you want to think about some of the implications of that.

Now, GPS does not only give position in terms of latitude and longitude on the Earth, or on a map that's built into a GPS receiver, it also tells you your altitude. It tells you how far above the ground you are, because it finds your position in three dimensions. It is therefore ideally suited to aircraft navigation.

In one system that went into effect in 2003, I believe, the wide area augmentation system is designed to bring the benefits of instrument landing systems to small airports that didn't have them before. Here's how it works. We've got a jetliner with GPS, and so, it's looking at some GPS satellites, and getting approximate information on its location, but not good enough to land with. It's good to within a few tens of feet. You might miss the runway, at that rate.

Deployed around the United States are 25 reference stations, which receive GPS signals; and they send those signals to central transmitting stations, which beam them up to a geosynchronous satellite. Remember that those are those satellites that are up 22,000 miles, which rotate once in 24 hours, and stay fixed above a given location on Earth. The geosynch satellite beams all of information back to the aircraft, and can tell the aircraft exactly where it is. At an airport that does not have instrument landing systems, this can take the place of instrument landing systems, and smaller airports can then be open to the kind of navigation that only larger airports had before.

By the way, another advantage of this is that it may save fuel, because it may allow flight controllers to vector aircraft paths that take them straight from one airport to another, instead of going in roundabout paths as required now by air-traffic control.

Well, there are many other uses for GPS in science, engineering, business, and so on. For example, if you're a scientist studying wildlife, you can stick GPS receivers on your wildlife. Let GPS satellites locate the wildlife, and then the transmitter on the little collar device that the elephant is wearing tells you, the biologist, just where the elephant is, and you can track it around. Oceanographers scatter GPS on the ocean surface, and they can use that to calculate the wind speed of the surface.

Engineers worrying about a dam breaking or bridges deforming, put GPS receivers on them, and they can actually measure the slight motions from that deformation. Geologists monitor the flow of glaciers, and even continental drift, the growth of mountains, with GPS.

GPS is everywhere. You may not realize it yet, but GPS is almost certainly part of your life, and it is going to become more a part of your life in the future. Just some examples: GPS tracking of your children, possibly, your pets, or your possessions. The GPS will come at some point, give us more intelligent cars and highways that may be able to drive themselves, because they will know precisely where they are, and where the cars around them are.

We will have improved GPS capabilities in the future consisting of greater power, more frequencies, more of them civilian accessible, more sophisticated timing, more pseudorandom codes, a link to the underlying radio signals, more power, and a whole bunch of new capabilities, with less interference. GPS is truly going to your part of your future, and is truly part of physics in your life.

Lecture Thirty-Two
Dance and Spin

Scope: Many objects, from electrons to dancers and skaters, CDs to
wheels, planets to stars and galaxies, undergo rotational motion.
Although such motion could be described in terms of Newton's
laws, it would be difficult to characterize the different speeds and
directions of all the parts of a rotating object. Consequently,
physicists describe rotational motion in terms uniquely suited to
the phenomenon of rotation. These terms are analogous to the
corresponding terms for ordinary straight-line motion, and a
corresponding set of laws describes them. But applied to rotation,
these laws lead to some surprising new phenomena that occur
throughout the realms of physics, including such everyday
applications as bicycle riding, ice skating, and weather. Rotation
can even provide a remedy for the apparent (but not actual!)
absence of gravity in orbiting spacecraft.

Outline

I.   When an object rotates, different parts of it move with different speeds
     and in different directions. In this situation, it's convenient to describe
     the rotation using quantities that apply to the entire rotating object.

   A. *Angular velocity* measures the rate at which an object rotates. As
        with ordinary velocity, it has a direction as well as a numerical
        value. That direction is given by curling the fingers of your right
        hand in the direction of the rotation; then your right thumb points
        in the direction of the angular velocity.

   B. Just as mass is a measure of how hard it is to change an object's
        motion, so *rotational inertia* measures how hard it is to change
        rotational motion. The rotational inertia of an object depends not
        only on its mass but also on how that mass is distributed relative to
        the axis of rotation.

   C. Just as momentum is the product of mass and velocity, so *angular
        momentum* is the product of rotational inertia and angular velocity.
        And just as the momentum of a system is conserved if no external
        forces act on it, so the angular momentum of an isolated system is

conserved. This fact has remarkable consequences from the atomic scale to everyday happenings to the cosmic realm.

1. Because angular momentum has direction as well as numerical value, the spin axis of an isolated spinning object will remain in a fixed orientation. This is the principle of the *gyroscope*, long used for navigation in vehicles ranging from submarines to ships to missiles. Gyroscopes are also used to stabilize systems that would otherwise be subject to excessive sway or vibration. Examples include cameras, telescopes, satellites, and cruise ships. A recent application is the use of micro-miniaturized gyroscopes in some GPS receivers (previous lecture!) to maintain location information if the GPS satellite signal is temporarily lost.

2. Change the rotational inertia of an isolated system, by redistributing its mass, and the rotational velocity must change to keep its angular momentum unchanged. Dancers and skaters routinely exploit this phenomenon to spin at high rates. Starting with arms extended, they go into a slow spin. Pulling in the arms greatly reduces the rotational inertia. To keep angular momentum unchanged, the rotation rate must increase. A cosmic application of this phenomenon is the *pulsar*, a rapidly spinning object with the mass of a star packed into a volume just a few miles across. Pulsars form when a star explodes as a supernova, collapsing much of the stellar material and, to conserve angular momentum, greatly increasing the spin rate. The fastest pulsars have been clocked at some 600 rotations per second.

3. Change the angular momentum in one part of a composite system, and other parts must respond to keep the original angular momentum constant. This phenomenon is used to steer the Hubble Space Telescope and many other satellites. Instead of using rockets, which would eventually run out of fuel and whose exhaust would interfere with Hubble's exquisite vision, the spacecraft includes three motorized wheels mounted in three perpendicular directions. Starting one wheel sets the telescope spinning in the opposite direction; stop the wheel and the telescope stops. In this way, the telescope can be made to point in any direction. Because the

motors run on electricity generated by solar panels, there's no fuel to run out.

II. The rotational analog of force is *torque*, which describes the effect of a force applied at some distance from a rotation axis. Just as force produces *change* in motion (Newton's second law), so torque produces *change* in rotational motion—specifically, in angular momentum.

A. Like other rotational quantities, the direction of torque is at right angles to the plane containing the actual rotation.

B. Understanding torque helps one appreciate the many skills that go into dancing.

　　1. Gravity exerts a downward force on a dancer, concentrated at the center of gravity. Unless the center of gravity is directly above the point of contact with the floor, that force will result in a torque that topples the dancer.

　　2. A solo dancer's feet provide the only place where forces may be exerted. Producing significant torque, especially *en pointe*, is particularly difficult.

C. Understanding the relation between torque and angular momentum helps explain some seemingly unintuitive phenomena associated with rotational motion. Remember that Newton's second law says that the direction of the force acting on an object is the direction of the *change* in the object's motion. Similarly, the direction of the torque on an object is the direction of the *change* in its angular momentum.

　　1. First consider a vertical object that isn't rotating, so it has no angular momentum. Like the dancer, if it's out of balance, there will be a torque that tends to topple it. While toppling, it is rotating about the point where it contacts the ground. Thus, it gains angular momentum. The change in its angular momentum is, indeed, in the same direction as the torque.

　　2. Next, consider what happens if the object is rotating. Now it has angular momentum, whose direction is along its rotation axis. If it is out of balance, there's still a torque, just as before, and that torque causes a change in angular momentum that's in the direction of the torque. But now the change adds to the existing angular momentum—and the effect is to change the direction of that angular momentum. The rotation axis

moves—in fact, it traces out a circle. This is the phenomenon of *precession*.

3. Precession isn't just about toy gyroscopes. It happens when any rotating object is subject to a torque. Our spinning Earth experiences a torque from the Sun's gravity, associated with the slight equatorial bulging of the planet. This results in Earth's rotation axis precessing once every 26,000 years. As a result, the star Polaris will not always be the North Star!

4. Precession even occurs on the subatomic scale. Subatomic particles, such as protons and electrons, possess an intrinsic angular momentum, called *spin*. Because they're electrically charged, this spin causes them to act like tiny magnets (recall Lecture Fifteen). An external magnetic field exerts a torque that tries to align the particles with the field. But because they're spinning, the particles precess rather than aligning. Precise measurement of the precession rate of protons in a strong magnetic field forms the basis of magnetic resonance imaging (MRI); more on this in Lecture Thirty-Five.

5. Finally, precession helps explain a common activity from everyday life—bicycling. What keeps a bicycle from falling over? If it's slightly off balance, there's a torque that would cause a stationary cycle to fall over. If the bicycle tilts to the left, that torque points toward the back of the bike, so the angular momentum of the bike's wheels must also change in the backward direction. That can be done by turning the wheel slightly to the left. This, instinctively, is what the rider does—and it's also what happens automatically, because of the torque, when one is riding using no hands! A moving bicycle is very stable because it is self-correcting—all the result of angular momentum and torque.

Suggested Reading:

W. Thomas Griffith, *The Physics of Everyday Phenomena: A Conceptual Introduction*, chapter 8.

Paul Hewitt, *Conceptual Physics*, chapter 8.

Going Deeper:

Richard Wolfson and Jay M. Pasachoff, *Physics for Scientists and Engineers*, chapters 12–13.

Questions to Consider:

1. When an ice skater spins rapidly by drawing in her arms, does a torque act to increase her spin rate? Does her angular momentum increase?

2. You can distinguish a raw egg from a hard-boiled one by rolling each down an incline, starting from rest. Which reaches the bottom first? Why?

Lecture Thirty-Two—Transcript
Dance and Spin

Welcome to Lecture Thirty-Two, "Dance and Spin." An alternative title might be, "Everything You Wanted to Know About Rotational Motion." This lecture is an outgrowth of Module Two, which dealt with going places, motion. This deals with a very specific type of motion, rotational motion about an axis.

Now, in principle, we can describe this kind of motion using Newton's laws, but the problem is, when we have a rotating object—like planet Earth, like a bicycle wheel, like as being a star, like a dancer doing a pirouette, like a proton spinning in the magnetic resonance imaging machine that's imaging your insides, all of these are examples of rotational motion—the problem is, different parts of the object are moving in different ways. They have different velocities. They have different directions of motion. They have different speeds, and it would be very complicated to describe rotational motion using Newton's laws, as we understand them.

That doesn't mean rotational motion isn't governed by Newton's laws, but what it does mean is that it is easier to come up with another language, an analogous language, to describe rotational motion. Ultimately, it is grounded in the physics of Newton's laws of motion, but this alternative language will make it much easier to talk about rotational motion.

I want to spend part of this lecture introducing that language, getting you to have faith that it is really exactly analogous to Newton's laws, therefore—what you already know of Newton's laws—being able to understand some phenomena of rotational motion, which are extraordinarily counterintuitive.

Let's get started. We would like to define analogues for the quantities that we know about, that define linear motion, straight-line motion, or motion in a circular path, something like that, but we want to define them for objects that are rotating. The first thing that we need is an analogue of velocity. That analogue is *angular velocity*.

When I talk about angular velocity—here's a little globe. I'm going to get spinning. That is what the Earth actually does. It has angular velocity. That part of the globe is moving that way. A part of the globe on the other side that you can see was moving that way. Different parts are moving in different directions. Parts down inside are moving with different speeds. We want to describe that with a single quantity.

I talk about how many revolutions per minute it's making, or how many revolutions per second, or how many degrees per second, or, more sophisticated, how many radians per second, or how many revolutions per hour, or in the case of the Earth, it undergoes the motion of one revolution per day, that's talking about something that all parts of that globe are doing. Remember that velocity has direction, and changes in direction count as much as do changes in speed.

The same is true for rotational motion, and again, that's a fact that is often overlooked. People who don't quite understand motion are forgetting that change in direction is a change in motion. Similarly, a change in the direction of angular velocity is also a change in its angular motion, its rotational motion.

How do we describe that motion in direction? Well, I'm going to be using my right hand a lot in this course, even though I'm left-handed. Here's way we do it. You might say, "What direction is this thing rotating in?" I could kind of talk about clockwise or counterclockwise, or I could sort of say, "Toward the east," but those are all with reference to the Earth itself. Here's the way to describe in good, solid, absolute terms.

What I do is to observe the object under rotational motion—in this case, the Earth spinning toward the east—and I curl my fingers of my right hand in the direction of that motion, this way. My right thumb points in the direction I'm going to define to be the direction of the rotation. Now, nothing is really happening in this direction. It's going around like this, but this direction, perpendicular to the plate in which the thing rotates, turns out to be the best way mathematically to describe the direction of that rotational motion. I'm therefore going to use this right hand rule.

If you're like me, and are left-handed, don't feel snubbed. We could have defined everything the other way. We would have had to draw all of our coordinate systems differently. The relation of magnetic and electric fields in a wave would have looked slightly different, but it all would have been described in the same physical world. However, we have chosen to use this right hand rule.

There, we have an analogy with regular, ordinary straight-line velocity. We have angular velocity. Now, we need an analogy with mass. Remember that mass is a measure of how heavy something is, or more importantly, how much resistance it presents to acceleration, to changes in motion.

Here's a very heavy thing. Uh! To get it accelerated, it takes a big force. Mass is a measure of inertia, resistance to change in motion. Well, in the case of rotation, the same thing is true. We have something analogous to mass that is called *rotational inertia*, and it's a measure of how hard it is to change an object's rotational motion. I have a picture here, three objects. They all have the same mass, the same resistance to changes in linear motion, straight-line motion, but they have very different resistance to changes in rotational motion.

The one at the upper left is like a bicycle wheel. It is mostly hollow. Most of its mass is concentrated in its rim. The one on the upper right is pretty much uniformly distributed mass, except for a little axle at the center. The one at the bottom has more of its mass at the center. Which one would be easiest to get rotating? Think about that for a minute. Well, the one with most of the mass at the center, because we don't have to get that mass up to very high speeds if it's close to the center, close to the rotation axis, but the bicycle wheel here has the greatest rotational inertia.

Just to give you a sense of how rotational inertia works, I have here two objects. They have identical radii—equal radius, equal diameter. They also have equal mass. I'm not going to weigh them. You will have to take my word for it. However, they have very different rotational inertias, because the distribution of mass in both of them is different. This ring has all of its mass at the outside, far from its rotation axis. This disc has its mass distributed uniformly.

I'm going to roll these two down this inclined plane, and you'll see quite clearly the fact that the rotational inertia of the brass ring is bigger than the rotational inertia of the wooden disk. It takes more—it doesn't respond—it doesn't change its rotational motion as rapidly as the wooden disk does, so here they go.

Let me do that again. Here they go. The wooden disk got to the bottom first, because its rotational inertia is less. It was easier to get it into rotational motion. Let me put these things away, because I will need this table for other things.

Now, we have a rotational analogue as well of mass. That's rotational inertia. In talking about linear motion, I defined *momentum*, quantity of motion, as the product of an object's velocity with its mass. Well, we can define the analogous quantity for rotational motion. The product of an object's mass analogue, rotational inertia, with its velocity analogue, rotational velocity, and that is called *angular momentum*. Just as the momentum of an object stays the same unless it is subject to forces—that is basically a statement of Newton's law—a statement of the analogue of Newton's law is that unless something is happening to an object—and we will talk about that something in just a minute—its angular momentum tends to stay the same.

That property is exploited, for example, in gyroscopes. I have a big model gyroscope. It consists of a big mass disk, mounted on a shaft. The whole thing is balanced very carefully. I have adjusted these weights to balance it, and I'm going to set the whole thing into vocational motion—or set the disk into rotation of motion. There goes. It's rotating; and if I let go, it basically stays exactly where it is. In fact, it's difficult for me to change the rotation of its orientation axis. Uh! It's hard for me to do that, because it wants to stay rotating in exactly the same direction. If I point it in the same direction, then the axis stays in that direction, because the direction of the angular momentum. In this case, the wheel is rotating that way, so if I curl my right fingers again that way, in that direction, along this shaft, it's the direction of this object's angular momentum.

If I represented that by an arrow, it would look something like this. The direction of the angular momentum of that rotating wheel is along the axis this way. If it were rotating the other way, it would have been done that way. That is angular momentum, and angular momentum tends to stay the same unless something—which I will describe in just a moment—acts on the object. This principle, by the way—the gyroscopic principle—is what is used, for example, to point spacecraft in space. It is used to stabilize ships. It is the principle of the so-called gyrocompass, because if I set a gyroscope in nice, frictionless bearings spinning, and keep it spinning, maybe by means of a motor that is mounted in the frictionless bearings as well, that gyroscope will always point in the same direction, even if the object around it—a ship, spacecraft, airplane, or whatever—changes the direction of its motion. The gyroscope will always point in the same direction. Why? Because its angular momentum doesn't easily change, and the bigger it's rotational inertia, the harder it is for that momentum to change.

By the way, I talked about GPS in the previous lecture. Some GPS receivers are coming out with tiny, motorized, rapidly spinning gyroscopes in them. Their purpose is to keep track of the GPS receiver's location in a brief interval when it might lose contact with satellites, as in going through a tunnel, or being hidden by the shadow of a mountain, or something like that. Gyroscopes then are used in navigation, in determining where things are, in determining changes in motion, because a gyroscope—just an object with a big rotational inertia, and therefore, a big angular momentum, if it's rotating fast—tends to keep its direction in space, because angular momentum doesn't easily change, just like regular linear momentum doesn't easily change, unless you exert big forces.

Now, there's an interesting difference between rotational inertia and mass. I can't very easily change the mass of my body. Well, I can change it gradually with my eating habits or my natural growth, but I can't easily reconfigure my body to have a different mass. I can, however, easily reconfigure it to have a different rotational inertia. Remember why the brass ring had a bigger rotational inertia than the wooden disk? It's because its mass was distributed differently. It had the same mass, but here, the mass was further from the rotation axis, and I am capable of rearranging my body in ways that alter my rotational inertia, and if I've got a certain amount of angular momentum, that angular momentum still can't change. It has to stay the same, and the angular momentum is the product of my rotational inertia and my angular velocity. If my rotational inertia goes down, my angular velocity has to go up, to keep my angular momentum the same.

Let me give you a quick demonstration of that, and to make it a little more dramatic, I'm going to hold these big weights out in my hand. I'm going to step on this turntable, which is fairly frictionless. It's not perfectly frictionless, but it's pretty good. I'm going to get myself rotating with these weights out, and I'm going to bring the weights in. I rotate faster. Here I go, round and round. I put the weights out, and I rotate much more slowly. What I am doing is making my rotational inertia bigger. I bring them in, and I rotate faster. I'm going to give myself a little more angular velocity there. Bring them in—whoo! Put them out.

I'm changing my rotational inertia. I'm therefore changing my angular velocity, but my angular momentum, the product of those two, stays exactly the same. Now, that principle, angular momentum staying the same as rotational inertia changes, is widely used in nature, technology, and in the arts, for example, and in athletics. Let me give you a few examples.

Here's an ice skater. It's Fumie Suguri from Japan. She's going to be performing on ice, and you will see her rearranging her body in ways that— at first, she has her body parts extended out from the axis about which she is rotating, and then, gradually, she's going to bring them in. As she does though, she is changing her rotational inertia, and therefore, she is changing the way her body rotates. She's changing her rotational velocity, but she's not changing her angular momentum.

There she goes. Watch her go around as she spins—there she goes—she stands taller and taller, and she still has to leg out, her arms out. Her rotational inertia is large. She is spinning rather slowly. Now, she's going to pull in her leg, pull in her arms. Faster and faster she goes; as her rotational inertia goes down, her angular velocity goes up, to keep her angular momentum the same. She is essentially spinning on frictionless ice, like I was standing on a frictionless platform there, and there she goes. By altering the configuration of her body, she's able to alter the rotational inertia, the resistance to changes in motion, but she can't change her angular momentum. Therefore, she has to go at a faster angular speed.

Now, there are many applications of this. There's a cosmic application. When stars explode, massive stars explode, the remnant of the massive star often collapses down to an object that is roughly a few miles across. A star might be spinning at a stately speed of the Sun's revolution, once per month, roughly. However, if that same object, with all that mass, collapses down to a few miles across, this principle of angular momentum staying the same, even as rotational inertia shrinks, means that those objects have to be going very, very, very fast.

Those objects are called *pulsars*, and we first discovered pulsars by the rapid "click-click-click" of the radio noise they made in our radio telescopes, because these objects were spinning, sometimes tens or hundreds of times per second. The pictures you see here are two pictures of the same object. It is the famous Crab Nebula, the result of a supernova explosion that took place—well, I shouldn't say that it took place 1,000 years ago, but it was first seen on Earth roughly 1000 years ago. The picture on the left was taken with an optical telescope. The picture on the right was taken with an X-ray telescope, and you can almost see something that looks like a rotating disk, and at the center of it is this pulsar. You can see two kinds of jets of material coming out either end. As this thing rotates, with different hotspots on it, and signals of beams of particles coming out, these give rise to radio signals that get brighter and dimmer, and brighter and dimmer, or louder and softer, as the pulsar goes around.

I have here some audio clips of this pulsar, the Crab Pulsar, which rotates at about 30 times per second, and another pulsar that rotates a little slower. Here's the Crab Pulsar, making that noise in a radio telescope. Here's another pulsar, called the Vela Pulsar, which is pulsing about 11 times per second. Those are the actual sounds of these rapidly spinning pulsars.

Now, we can change a system by changing its rotational inertia. We can also change a system that has several different parts, by changing the angular momentum of different parts of that system. Let me give you an example of that.

Here I have this bicycle wheel I've talked about several times before. I'm going to stand on this platform. Now, right now, neither I nor the bicycle wheel are rotating—very much, anyway—and so, we have no angular momentum whatsoever this system has no angular momentum, zero. However, I'm going to get the bicycle wheel rotating. Well, the system still has no angular momentum, but the bicycle wheel was rotating in one particular direction. Remember that angular momentum has direction, and in fact, I set it rotating that way, so it's angular momentum was up. In order to conserve angular momentum, to keep it the same, still zero, I had to rotate it the other way, net angular momentum down, to keep the angular momentum zero.

I want to emphasize that important point. Angular momentum has direction, and a change in directions is as much a change in angular momentum as is a change in speed. That was the key to understanding a lot of things like satellite orbits and so on, in the Newtonian case, and linear motion. It's going to be the key to understanding some really weird phenomena, as well as some common ones—like riding a bicycle in the rotational case.

Let me do another example. Now, I'm going to give the bicycle wheel some angular momentum, and I'm going to step onto the wheel. Now, the system has angular momentum, and I'm going to turn the bicycle wheel upside down, which dramatically changes in angular momentum, because it changed the direction of that angular momentum, and the whole system still has the same angular momentum that it did before, but now, the two parts of that system have different angular momentums that conspire to give what was the initial angular momentum of that system. That principle, by the way, is put to very good use in spacecraft, particularly the Hubble space telescope. Many spacecraft are actually steered—pointed at the objects they're looking at, if they're astronomical spacecraft—by means of angular momentum.

We could do it with little rocket jets, but sooner or later, we would run out of fuel, and in the case of Hubble, we would have all kinds of rocket exhaust around, blurring the images, so what we have instead in a spacecraft like Hubble are four so-called *reaction wheels*. They're like my bicycle wheel, and they can be spun up or down. They are mounted at right angles to each other, and there's a fourth one at an angle for redundancy, in case one fails. It would take three at right angles to everything we need to do. They're spun by motors, which are powered by solar panels on the spacecraft, so that they can't run out of fuel. When we want to spin the spacecraft in one direction, we rotate one of these wheels in the opposite direction. By having wheels in three different directions, we can do this in any direction. That's a real practical use of this phenomenon.

Well, we now have the rotational analogies of a great many quantities, but we haven't yet got the rotational analogue of force. I want to give you a sense of what that is. Here's the bicycle wheel. If I want to change its rotational motion, I have to apply a force. That force is more effective if I apply it there, than it is if I apply it there. When you try to open a door, you make the door rotate. You don't push on it near the hinges. You push on it on the far side. That's where we put the handle. Why? Because it's more effective at changing rotational motion.

That quantity that changes rotational motion is called *torque*. Torque is a combination of force, and the distance the force is from where the rotation is occurring. For example, here's a bolt stuck into a great hunk of metal. I want to turn that bolt—uh! I can't do it with this little tiny wrench. However, if I put on this much bigger wrench, where I can apply a force farther away from the rotation axis, then I can easily tighten the bolt. That quantity torque—the analogue of force—depends on the force, and how far you are from the axis of rotation. Here's a picture of a wrench. I want to rotate it in what in this picture is a clockwise direction. I could apply a force some distance from the rotation axis. I wouldn't get much torque. It would not be very effective.

I could apply a force like that, but because it's pulling away, partly, it wouldn't be very effective either. I get the most effect, the biggest torque, if I apply it at right angles like that, and at a long distance the rotation axis. Thus, now we know the rotational analogue of torque as well. What's the direction of that torque? The same thing. Right hand rule, twist in the direction the bolt would turn. It's into your screen, in this particular case. There it is.

Here are analogies: Velocity, "V," and angular velocity is given the symbol of Omega, but we won't worry about that. Mass, "M," rotational inertia—momentum, "MV," angular momentum, the product of rotational inertia and angular velocity, force, torque. On the basis of those analogies, we are going to understand some pretty strange things that happen with rotational motion. Let's look at some simpler things first.

First, we have to understand the analogy with Newton's second law, and that's another complete analogy once we have this picture. Newton's second law says that force is the rate of change of momentum. The rotational analogue of Newton's second law says that torque is the rate of change of angular momentum. That's all we need to know.

Let's look at a few examples from dancing, for example. Here's a dancer, balancing. Her center of gravity is right over the point of contact with the floor, her foot, and there are no torques exerted on her. If she leans over little bit, the force of gravity is straight down. If she leans over a little bit, the force of gravity is still straight down, but now, there's an angle between her foot, and that center of gravity, center of mass, and she's going to topple over. There's going to be a torque on her. The torque is going to be—again, applying the right hand rule in this way—she's going to tend to topple to what is the right on this screen. That's going to be an angular momentum into your screen—well, it is going to give her an angular momentum into your screen. It's going to be a torque into your screen.

Here's a dancer trying to pirouette. It's a very difficult thing to do, especially if the dancer is on her toes. If you look closely at her foot, there's only a tiny little bit of foot in contact with the floor. How does she apply torque to keep herself going? Well, if you look carefully at her, she drops down onto her flat foot, and she's able to apply, between the heel and the toe, or the heel and the ball of the foot, forces in slightly different directions, and they result in that torque that actually spins her around.

Here we have Candice Ulbrich, whom we saw way back in Module Two, from the Loudoun Ballet Company. What she's going to do for us is a *fouetté* turn. Watch her carefully. You'll see her spinning around. Very difficult to get any torque when she's on her toes, but then when she goes down onto her flat foot, you'll see her exerting the forces on the floor that result in the torque, that gets her spinning. There she goes, doing a turn. Notice, as well, her leg is going in and out. She's changing her rotational inertia as she does this. There's the toe. Very difficult with the toe, but when her foot goes down, that's the time she gets the torque. There she is, spinning for us. Thank you.

Finally, there's a very strange phenomenon that occurs with rotational motion, which you may find baffling. I want to show you how that works. If you understand that torque causes changes in angular momentum, you'll understand this. First, consider an object, like this meter stick that is not rotating. Suppose it starts to fall over. Well, like that dancer, there's torque on it. There it goes.

It has started rotating in that direction. It's gained angular momentum, which is that way, the way my thumb is pointing. There's a torque on it in that direction, and everybody is completely consistent. If I had a gyroscope, a little thing that spins that was not rotating, and I have it unbalanced, like this, sitting on a little pedestal—there's its pivot point; there's the force of gravity on it. There's a torque on it, and again, right hand rule—there's the direction of the torque, into your screen.

What's the direction of the change in angular momentum? Well, change in angular momentum, torque. The same. What's the direction of the force? The same as the direction of the change in an object's motion. What's the change in direction of the object's angular motion? The same as the direction of the torque into the screen.

There's no rotation, so the angular momentum, which was zero, has to increase, and the gyroscope falls over. The direction is that way, into the screen, and clockwise rotation. The gyroscope falls over.

What happens if the gyroscope is already rotating? If it's already rotating, then it still has a torque on it. The direction of the torque is still the same. It's into your screen. By the way, if this is getting complicated, play it over a few times. It's into your screen, but now the rotating gyroscope already has angular momentum, along that access, it is rotating in the direction I've shown.

What has to happen? Well, what's the direction of the change in angular momentum? Again, it's the same as the direction of the torque, and that is again into your screen, but because the gyroscope is already rotating, just like satellite is already moving—it doesn't fall to Earth; it falls around Earth—the gyroscope undergoes this processional motion, in which it rotates around like that.

Let me give you an example of that with the gyroscope I have over here. Here's the gyroscope. It's perfectly balanced right now. I'm going to wind this cord up on it, so that I can get it rotating again, just like you would do with a little toy gyroscope. I've got a little knot there, but I won't worry about that.

Okay, there it goes. It's rotating happily. Its angular momentum is staying constant. It's balanced. Now, I'm going to imbalance it, and that will put a torque on it. Now, its angular momentum changes, in the direction of that torque. There it goes, happily around, in the direction of the torque. The angular momentum is changing in this direction, and that is the direction of the torque.

Let me give you another example of that. Here's my big, gyroscopic bicycle wheel. This is not actually a bicycle wheel. This is a bicycle wheel that has had a much more massive rim. Instead of an air-filled tire, there's a lead-filled rim here, covered by rubber—lead, or some heavy thing. This particular bicycle wheel has a very large rotational inertia, much larger than what you would want to have on your bicycle. I'm going to take this bicycle wheel, and I'm going to put a rope on it.

Now, if I simply dangle it from this rope, it will, of course, fall over. No question. However, what if I get it rotating? Well, now it's already got angular momentum. I hold it unbalanced, and there it goes, processing around again. The torque on it, which would make it fall over if it weren't rotating, is doing the same thing, changing its angular momentum, at the rate given by the torque. That strange phenomenon of procession is really just a consequence of Newton's second law, applied to rotational motion.

There are some practical examples of that. Earth, for example, is itself a big gyroscope. It's a big spinning thing. Because of its rotation, it's not exactly a sphere. In fact, it's wider at the equator, so it looks something like this, in exaggeration. It's spinning around. Right now, its polar axis points to the star Polaris. There's a force of the Sun's gravity on one side of the Earth, and there's a weaker force of the Sun's gravity on the other side of the Earth, because gravity falls off with distance. Together, they conspire to make a torque on the Earth, and that torque makes the Earth process just like a giant top. As a result, 13,000 years from now, the star that is Polaris will not be the North Star. In fact, the star Vega will approximately be the North Star. That procession occurs on Earth with a procession of about 26,000 years. Twenty-six thousand years from now, Polaris will again be the North Star.

By the way, I mentioned in the discussion of climate that there were some changes in Earth's orbit that will affect climate. This is an example of one of those kinds of subtle changes that can affect Earth's climate.

Procession also occurs on the subatomic scale. Here's a proton. Protons have a little intrinsic angular momentum called *spin*. This proton is spinning around. It's also got a north and south magnetic pole. It's magnetic, because it's charged. The charge is moving, and that makes magnetism. If I put the proton in a magnetic field, there will be torque on it, and the direction of the torque—in this case, given the direction I've drawn the spin, spinning that way—with force tending to bring it toward alignment with that magnetic field, so that it's going to tend to rotate that way. That's going to make it torque out of the screen, and consequently, it's going to process in this direction. The pole is first going to move out of your screen, and then back around into your screen.

That is not an unimportant phenomenon. That phenomenon is the basis of magnetic resonance imaging, which I will talk about more in a few lectures, in the lecture on physics in your body.

Finally, let me use these ideas to explain a much more common, everyday experience, which it really was not until about the 1970s, when we really began to understand fairly well how a bicycle works. What keeps a bicycle from falling over? What makes a bicycle so easy to ride?

Well, I've got a bicycle here. Now, here's a bicycle, and if it is standing still, and it starts to fall over this way, to the left, it will simply fall over. Well, if it's like this, there will be a force on it downward. That will result in a torque. Take my right hand, going the direction it's going to rotate, backwards, so that if the bicycle tips to the left, there is a torque toward the back. That means that it angular momentum has to change in the direction toward the back. Now, if the wheels are not rotating, it has no angular momentum, so it's got to gain angular momentum. It has to go from zero angular momentum, to an angular momentum that points toward the back. That can happen if the whole bicycle rotates, tipping over, like that, because it gives an angular momentum to the back.

What if the wheel is already rotating, though, like that, in the forward direction? Well, then the bike has angular momentum that way, right out of the axle, and I'm going to stick a little arrow on here to remind you about that angular momentum. There's the direction, then, of the angular momentum of the bicycle, when the bicycle wheel is rotating, and the bicycle wheel is going forward.

Well, if the bicycle starts to tip over, all the arguments I just made about the torque are still there. There's now a torque on the bicycle. The direction of that torque is the direction in which this rotation wants to occur, if the bicycle weren't rotating, namely, still toward the back. None of that argument changes. Now, though, what does that mean? It means that the angular momentum has to change, and has to go more toward the back.

However, the angular momentum is already in that direction, essentially horizontal, so how does that direction changed toward the back? Well, easily. Just turn the wheels a little bit. Then, Newton's second law, the analogy for rotational motion, is completely satisfied. The way you ride a bicycle, and the reason that he so easy to ride a bicycle and stay balanced is—this is at least part of the reason. There are some other things at work here, too—is that the angular momentum change that is necessary in response to that torque that tends to pull you over is simply a slight turn of the wheel. As a result, bicycling becomes very simple.

Lecture Thirty-Three
The Light Fantastic

Scope: The laser is among the most important inventions of the 20th century. Superficially, the laser's extraordinary intensity and sharply defined beam are what distinguish it from other light sources, but the distinction is actually much deeper. Unlike other light sources, lasers exploit a remarkable form of cooperation among atoms, a cooperation mediated by light itself. The result is a light beam that is, most importantly, coherent—meaning that the light waves emitted by individual atoms are all precisely in step. This coherence is ultimately responsible for laser light's more obvious properties, including sharp beams, high intensity, and purity of color. A wide range of substances—gases, liquids, and solids—can be made to produce laser light. Laser applications range from the obvious and everyday, such as bar-code scanners and laser pointers; to less obvious but still everyday applications, such as sending email and computer data over the optical fibers of the Internet; to industrial applications, such as welding or cutting steel; medical uses, including laser surgery, cancer treatment, and diagnostic microscopy; surveying, ranging, and construction applications; and a host of others. Laser technology is indispensable in the modern world.

Outline

I.   Light is most commonly produced when electrons in atoms jump from higher to lower energy levels. Light or other forms of electromagnetic radiation can also be emitted by jumps among energy levels in molecules, in atomic nuclei, or in semiconductors (recall Lecture Nineteen).

    A. In a typical light source, such as an ordinary lightbulb, a heated object (the lightbulb's filament) gives off light. This occurs because atoms in the hot object jostle violently against each other, exciting atomic electrons to higher energy levels. Very quickly, the excited electrons jump back to lower energy states, giving up their excess energy in a burst of light. This process is called *spontaneous emission*, and it happens at random in a heated object, with light emitted in all directions. Light from the Sun arises in essentially the same way as in a lightbulb.

B. In fluorescent lights and neon signs, an electric current excites atomic electrons in a rarefied gas. Again, the electrons drop to lower energy levels, emitting light in the process. Because the gas is rarefied, the atoms don't interact and a spectrum of discrete colors is emitted, with wavelengths related to the energy difference between the two levels. But again, the process is spontaneous; there is no relation among the bursts of light emitted by different atoms.

C. Spontaneous emission is not the only reason an atomic electron will drop to a lower energy level. In 1917, Albert Einstein recognized that an electron at a higher energy level could be stimulated to drop to a lower level by the passing nearby of light that had been emitted earlier by the dropping of an electron in a similar atom. This *stimulated emission* is the basis of laser operation.

II. *Laser* stands for Light Amplification by Stimulated Emission of Radiation.

A. The key to laser action is to prepare a large number of atoms with electrons in excited states. Normally, most atomic electrons are in the lowest possible energy state, but in a laser, that situation is reversed; therefore, the situation is called a *population inversion*. The atoms can be excited by light, by an electric current, by chemical reactions, or other means.

B. The first working laser, built in 1960, used a ruby rod surrounded by a flash lamp similar to a camera flash. Light from the flash excited atoms in the ruby, providing the population inversion. Today's lasers use solids, liquids, or gases as the excited medium.

C. To make a laser, the substance with excited atoms is surrounded by two perfectly parallel mirrors, one of which is not quite perfectly reflecting. When an excited atom drops to its lower energy state, it emits a burst of light. Passing another exited atom, the first burst stimulates a second identical burst—identical in phase (its wave crests align with those of the first burst) and the direction in which it's going. Some light is lost out the sides of the device, but light that happens to be traveling perpendicular to the mirrors gets reflected back and forth, making more and more stimulated emission and building up a significant light intensity between the mirrors. A tiny fraction of this light escapes through the imperfect mirror to make up the laser beam. The beam is remarkably coherent, with waves remaining in step over distances of typically 20 centimeters (about 8 inches), or about half a million wave cycles.

D. Lasers can also make use of excited states in molecules or in semiconductors. Semiconductor lasers, also called *diode lasers*, are now very inexpensive and widely used. A semiconductor laser consists of a junction between P- and N-type semiconductors (recall Lectures Nineteen and Twenty), with highly polished edges that act like mirrors. An electric current flowing through the device creates electron-hole pairs, which recombine at the junction, emitting light. The emitted light stimulates additional stimulated electron-hole recombination, resulting in laser action. The color or wavelength of the light is determined by the semiconductor's energy band gap (recall Lecture Nineteen).

E. Lasers are available with power outputs from less than a milliwatt (1/1000 of a watt) to millions of watts. The highest power lasers produce very short bursts of light with peak power as high as 1,000 trillion watts.

F. Lasers are available with "light" wavelength or color across most of the electromagnetic spectrum, from microwaves to ultraviolet. Infrared and visible-light lasers are most common. Some lasers are tunable, with output wavelength variable over a wide output range. X-ray and gamma-ray lasers are under development. Very high power x-ray laser action has been achieved using nuclear explosions as the energy source or pump.

III. Today, lasers have almost unlimited uses in consumer products, commerce, industry, and science. Here is just a sample from those many uses:

A. In communications, diode lasers incorporated into integrated circuits convert electrical signals into light for transmission along optical fibers. Most computer networks are based on such optical communication; so are many land-line telephone systems.

B. Lasers are at the heart of optical information storage systems, including CDs and DVDs (Lecture Two). Low-power lasers "read" the information on discs, and higher power lasers "burn" discs to store information (Lecture Two). The development of higher energy band-gap semiconductors enabled the higher information capacity of DVD, and progress continues with the new "blu-ray" discs and blue lasers that read them.

C. Lasers "read" the bar codes at retail and grocery store checkout counters, on library books, on airline tickets, and on shipping and tracking labels.

D. Lasers with power outputs in the kilowatt (1,000 watts) range make precise cuts in steel and other metals and can weld metals together. Laser cutting tools can be programmed to cut arbitrary shapes, saving manufacturers the cost of developing custom tools for new product models. Lower power lasers (around 100 watts) are used for engraving, shaping "laser cut" keys, and similar applications. In a related application, lasers are used to harden metallic surfaces, resulting in longer wearing parts for machinery, cutting tools, and other uses.

E. Laser beams can destroy biological tissue, allowing precision laser surgery. Laser beams can be guided through optical fibers deep into body cavities, making laser "knives" less invasive than conventional surgery. Laser vision correction uses this process to reshape the cornea (Lecture Four). Laser beams can also "weld" detached retinas to the correct position at the back of the eye. More on these topics in the next lecture.

F. Precise, high-speed steering of laser beams by optical or mechanical means allows laser beams to form or reproduce images. Applications include laser printers and digital copiers, huge (26-foot!) television screens, and laser light shows.

G. Lasers are widely used in pollution analysis and other applications requiring remote sensing of the environment. Tunable lasers (variable color, wavelength) can be set to the precise wavelength that a particular substance absorbs. By observing scattered light, the type and amount of different contaminants may be determined. This technique may prove useful in detecting minute quantities of materials escaping from clandestine nuclear weapons construction, thereby helping to slow nuclear proliferation.

H. Laser beams can trap tiny objects, making possible "optical tweezers" that manipulate microscopic structures in individual cells.

I. Timing of reflected laser pulses provides precise distance and speed measurements. Terrestrial uses include everyday surveying, assembly-line positioning, missile guidance, and law-enforcement applications, such as an automated system for catching red-light runners. In space, lasers track satellites to provide precise position and orbit measurements. Laser beams sent from Earth to reflectors left on the Moon by Apollo astronauts provide measurements of the Moon's position to better than an inch and help verify Einstein's general theory of relativity.

J. The straightness of laser beams makes them useful for leveling, alignment, squaring, and similar procedures in construction and manufacturing.

K. Very brief laser pulses (trillionths of a second or shorter) are used as high-speed flashbulbs to "freeze" steps in complex biochemical reactions. Much of what we know about the details of such reactions comes from this technique.

L. In physics research, lasers provide a means of cooling to very low temperatures (millionths of a degree above absolute zero). A laser beam is tuned to the precise wavelength that can be absorbed by atoms in thermal motion. When an atom moving toward the laser absorbs light, the atom slows. The effect on a group of atoms is to lower the temperature.

M. Coherence of laser beams makes wave interference easy to detect with laser light. Interference-based instruments allow precise measurement of distances with accuracies of a fraction of the wavelength of light (less than a millionth of a meter). Such devices range from industrial instruments that accurately measure the deposition of materials in semiconductor manufacture to a proposed space-based system beaming lasers around a triangle 3 million miles on a side and intended to detect elusive waves of gravity.

Suggested Reading:

Louis A. Bloomfield, *How Things Work: The Physics of Everyday Life*, pp. 546–550.

Marshall Brain, *Marshall Brain's How Stuff Works*, pp. 83–84.

Paul Hewitt, *Conceptual Physics*, pp. 597–601.

Going Deeper:

Theodore Maiman, *The Laser Odyssey*.

Charles H. Townes, *How the Laser Happened: Adventures of a Scientist*.

Richard Wolfson and Jay M. Pasachoff, *Physics for Scientists and Engineers*, pp. 1125–1127.

Questions to Consider:

1. A laser pointer has a power of only 5/1000 of a watt. Yet its spot on the wall is brighter than the illumination provided by a 100-watt lightbulb or even a 500-watt spotlight. Why?

2. What is a population inversion, and why is it critical to laser operation?

3. Name the most recent encounter you've had with laser technology.

Lecture Thirty-Three—Transcript
The Light Fantastic

Welcome to Lecture Thirty-Three, "The Light Fantastic." Surely, the laser ranks right up there with the transistor as one of the most important inventions of the 20th century. You probably think of a laser as something of a light source, like a light bulb, except that it produces a very bright spot. There it is, on the screen over there. A very bright spot, very intense, and it tends to be a very narrow beam.

Those are, in fact, distinguishing characteristics of laser light, but the difference between lasers and other light sources go much, much deeper than that. Lasers would be a scientific curiosity, or maybe a scientific instrument for use in research, but for the fact that they're widely used throughout commerce, industry, science, medicine, and almost every other aspect of our lives.

I have a couple of pictures here on the screen, just to remind you of that. On the left, you see a laser system being used to scan the barcode on a carton of milk. On the right, surgeons are using laser light to perform delicate eye surgery. Lasers are very much a part of our lives.

In this lecture, I want to give you a sense of how lasers work, why they're so dramatically different from other light sources, and then, give you a panoply of applications to which we put laser light.

Most light that we see in the world originates in atomic scale systems. I argued that when I talked about electromagnetic waves, and how the size scale that produces electromagnetic waves is somehow related to the wavelength. It happens that atomic-sized systems typically produce visible light. A quantum mechanical way to put that is that the energy levels in an atom are typically separated by energies that correspond to the little minuscule units, the little *quanta*, the photons, that make up visible light.

I want to explain lasers in the context of that atomic physics and the transitions that electrons go through in atoms. You remember that we've seen this picture of an atom several times before. It's the sort of solar system model. It isn't exactly correct quantum mechanically, because the electrons are really in a kind of statistical cloud, but it's not that either, and what it shows quantum mechanically are these discrete levels in which the electrons can "live," in which they orbit the nucleus. Again, they aren't really nice circular orbits like that. They aren't really orbits at all. The electrons are distributed statistically, but this model tells us a lot about how atom's work.

There are therefore these atomic energy levels, and we often represent the energy levels as with a diagram like this, and you have seen this before in the context of our discussion on how semiconductors work from a quantum mechanical point of view. Thus, this particular atom on the left has three electrons. Two of them are in the innermost, lowermost, lowest energy orbit, and there is one electron all by itself in the middle orbit. The outermost orbit has no atoms in it. It's empty, and it's possible to promote that outermost electron to a higher energy level. One way to make that happen is to have a burst of electromagnetic energy come along with just the right energy to give the electron the energy to make that transition.

Now, I haven't said a lot about quantum physics in this particular course. You can learn more about it in my course on modern physics for nonscientists, but the fact is that light, in addition to being an electromagnetic wave, also consists of particles. I'm not going to get into the seeming contradiction there, other than to say it's not a contradiction.

The sense for us in which light consists of particles is that light energy, instead of being spread equally over the wave, is, in fact, quantized in little tiny bundles of energy, with discrete values that depend on the color of the light, and you can't have any less energy than that.

Here comes a photon then, which I will represent as a little wavy thing, because it also has a wave characteristic to it. That photon gives up its energy to the electron, and the electron jumps to the higher-level state. I've shown that happening in the picture of the atom, and in the energy level diagram at the left. Understanding that process—jumping up and then jumping down again, and emitting a photon of exactly the same energy, of exactly the same color of light—is what goes on when atoms either absorb or emit light.

We want to simplify this picture. I don't want to be drawing atoms all the time when explaining how lasers work, so a simplified picture of an atom conceptually would be just this energy level diagram. It can get even simpler than that though, because we aren't really concerned about what those innermost electrons are doing. They're just sitting happily in there over it, and aren't jumping up and down. It's only the outermost electron that does so, and so my simple representation of an atom from now on, in this lecture, is going to be a few horizontal lines that represent separate energy levels, and a blue circle that represents an electron, that can be in one of those two levels. There, then, is our very simple picture of an atom, and every time you see that, think "atom."

Now, here are three atoms. Two of them happen to be in that excited, higher energy state, and they tend to go back down naturally, spontaneously, to the so-called *lower-level* states, sometimes called the *ground state*. Why does that happen? Well, it happens for the same reason that if I put a ball on the table, sooner or later, it's likely to fall off. There's the probability that that electron will make that transition downward. It happens for atoms in very short times, way under one millionth of a second, or one billionth of a second.

Therefore, if an atom gets excited, it spontaneously falls back to the lower energy level, and emits a photon. That atom has just done so, and then, a while later, maybe that atom does so, and the photons they emit are going in different directions. They may have exactly the same color, the same wavelength, because it's the same energy transition, but they're going in different directions, and more importantly, the waves are not in step. They are what is called *incoherent*. The crest of one wave does not line up with the crest of another wave, and the key to laser light—the thing that makes laser light different from other kinds of light—the key to that bright, intense spot of laser light, is not the brightness, intensity, and narrowness of the beam, but the fact that all the photons, all the waves, if you will, that make up that laser beam, are coherent. They're all in step. That's the key to laser action.

If we have regular incandescent light bulbs, they are emitting light by just the random jostling of the rapidly moving atoms in the hot filament. A regular, ordinary light bulb emits light of all different colors, as atoms are getting jostled. Their energy levels are getting squeezed around. There are all kinds of different energy levels that get different colors out, and they're completely incoherent. That's the difference between a light bulb and a laser. It emits incoherent light, random phases, random times when waves are at their crest. With a laser, the wave crests are basically all coherent.

Even in lights like fluorescent lights and neon signs, there's an electric current there that is exciting atomic electrons, and they tend to emit discrete colors of light, although in a fluorescent lamp, they're changed from ultraviolet to visible by the phosphor coating on the inside of the lamp. In a neon lamp, you're actually seeing those discrete colors. Still though, even though the colors are discrete—because they are due to particular atomic transitions from one energy level to another—nevertheless, the light coming out isn't coherent, and so, it's not laser light.

Now, spontaneous emissions—this sense that if I get an electron up to a higher energy level, it will spontaneously drop—are not the only reason an electron can drop. In 1917, Einstein, in one of his many brilliant realizations, recognized that if you had an electron at a higher energy level, and a photon came by that had the right energy to have pushed that energy up to that level, that photon could also stimulate the electron to drop back down. Thus, ironically, a photon that could have put that electron up into that level, if it is already up there, can stimulate it to drop back down. That's called *stimulated emission*. That's the "SE" in the word "laser," by the way.

What happens? A photon comes along, it has just the right energy corresponding to that transition—it doesn't, this won't happen—and it goes by, the electron drops down, and it, too, emits a photon. Now, we have two photons, and in the process of stimulated emission, those two photons are moving in the same direction, and they're exactly in phase. In a quantum mechanical sense, they are identical particles. They're moving identically. They've got the same phase. Their crests all line up, and they are perfectly coherent. The original photon and the new photon—there they are, crest lining up with crest, trough lining up with trough, and so on.

That's the process of stimulated emission. What "laser" stands for is "Light Amplification by Stimulated Emission of Radiation." Light amplification—a laser is a kind of amplifier, as you will see in a minute. It amplifies a weak level of light. Here's how it does it. If one photon comes along, that's a pretty weak light. Pretty soon, you have two photons. Pretty soon, you have more photons, and more photons. Light amplification by stimulated—that's this process—emission of radiation. There are also *masers*, which are "Microwave Amplifiers by Stimulated Emission of Radiation." In fact, they preceded lasers. That's what a laser stands for.

How do we make a laser? Well, the key to laser action is to produce a large number of atoms in this excited state. By the way, they don't have to be atoms. We can also excite molecules; we can excite electron hole-pairs, and semiconductors. There is a wide, amazing diversity of ways to make laser action happen, and we have explored many of them. However, all of them involve creating what is called *population inversion*, a situation that the material the laser is made of doesn't normally find itself in. I'm going to think of our laser as being made of atoms of gas, and that is a common kind of laser, but it is certainly not the only kind of laser.

Here we have these atoms then, and normally they're all in their lowest energy state. Once in awhile, one or two of them get knocked up to a higher state, but that's just a few of them. In the laser, you have to do something. You have to supply energy in a process that's called *pumping*, that puts a large number of atoms—or whatever the lasing medium is—in this higher energy state.

The first step is to establish what is called a "population inversion." Here, again, is a picture of three atoms. This is my simplified picture of an atom, and they're all in that higher energy state. I have established a population inversion. There are many ways to do that.

The very first laser, in fact, used a crystal of ruby, a cylindrical crystal of ruby as its *lasing medium*, as it's called, and wrapped around it was a xenon flash lamp, one of those bright white lamps like you use in a camera flash, but wrapped around in the form of a tube, and when the xenon flash lamp went off, it gave off a lot of photons of different energies. Some of them were right to stimulate the atoms in the ruby crystal. That's what created population inversion.

Once you have the population inversion, one of those electrons will spontaneously drop back down to its lower energy state. All around it though, are other electrons in this excited state, and now, the process of stimulated emission occurs, so the proton from the leftmost electron dropping came by the middle one, and stimulated emission, and those two photons going happily by the next one stimulate an emission there. Now, we have three photons, all completely coherent, all completely in step, representing light, and it is now three times more intense than it was with that original spontaneous emission. When we have lasers, therefore, we are amplifying the initial natural spontaneous emission that occurs by this process of stimulating emission, and this happens because we have somehow artificially done what nature doesn't usually do; that is, to create a population inversion, a situation where we have a lot of atoms in the excited state, rather than the lower energy state.

Now, how do we make a complete laser out of this? Well, there's a little bit more that we have to do. Again, here's my symbolic picture of a bunch of atoms, represented as little horizontal lines. Most of them are in the excited state. One or two of them here are in the lower energy state, but most of them are excited. The population inversion isn't necessarily complete. Out of the system, due to light amplification by stimulated emission of radiation, comes a bunch of photons, all completely coherent.

Now, just a blob of gas, put into a population inversion state, is not enough to make a laser. What we do as well is to add mirrors on either side of this region containing gas, this piece of ruby, or whatever we've got. We put mirrors, and notice that these two mirrors are different. One of them, on the left, is darker. That mirror is an almost perfect mirror. It reflects 99.9999999% of the light incident on it. The mirror at the right isn't quite perfect. It might reflect 99%, or 99.5%, or 99.9%, of the light coming on it, but it is specifically designed so that a little bit of the light incident on it leaks out.

What happens? Well, we've got a whole bunch of these photons, and they've come, and they've hit that not-quite-perfect mirror, but most of them have reflected off, because it's almost perfect. They were produced by that initial spontaneous emission, and then by the stimulated emission that I showed you in the previous picture, they go back through the medium again, they stimulate more stimulated emission, because not every electron has dropped back down again. They bounce off the other mirror, they go back through, and they stimulate still more, and a little tiny fraction of that energy in the form of light leaks out through that little imperfect mirror, and that is the laser beam.

Thus, when I turn on my laser here, and shine it on the wall, that beam is just a tiny, tiny faint residue of the extraordinarily intense laser light that is bouncing back and forth inside the laser itself between those mirrors, light that I could never get access to, because if I tried to, I would break the laser action. That light, though, is much more intense inside the laser. It's only a little bit that leaks out, to make the nevertheless very intense laser beam.

By the way, you might say, "What happens when these upper levels are depopulated?" Well, lasers can work in one of two ways. You can fire a so-called *pulse laser*—some of the highest-powered lasers are pulse lasers. You build up the population inversion, the laser fires, and makes an intense but short burst of light, and then, you spend some time rebuilding the population inversion.

More common lasers, *continuous lasers*, are on continuously, and the pumping is happening continuously, even as the device is lasing—that is, producing laser light. For example, this is a helium neon laser. It has a tube containing helium and neon gas, and an electric current is being driven through that. That's what causes the electrons to the excited, and drives the population inversion at the same time. When I turn it on, it's lasing.

Over here, I have an example of laser, and this one is specifically designed to be visible. What you see are some electronics that govern the pumping action, and so on, and some other aspects of this particular laser I will get to later. Here's a laser too, though. It's a glass tube. It is actually not this big tube, about an inch in diameter, but a smaller tube down inside that, that you can see if you look close up, and when I turn the laser on, that small tube glows very vividly, just like a neon lamp would. In fact, that's what it's being. This is a helium neon laser, so it's glowing with the colors of a helium neon laser, and some other colors, too.

The light you are seeing, the intense pinkish light that is coming out of there, is not the laser light. Some of the photons that are emitted don't go out in the direction parallel to the laser beam, and they're lost, so the laser is not perfectly efficient. There's quite a bit of light being produced that's going out in different directions, is not going along bouncing back and forth between the mirrors, is not stimulating more emission, and is not contributing to the laser action.

Quite a bit of the electrical energy then is producing that glow, which is not doing anything for us. It's only the photons that happened to be emitted, going back and forth, that build up more and more levels, and they emerge as the laser beam, which you can see here on my hand. Again then, that's a tiny residue of what is bouncing back and forth inside that laser cavity, between two mirrors—the partly silvered mirror here at this end.

That's what's going on inside the laser. It's not a light bulb. It's not even a neon lamp, although it has that same glow the neon lamp does. The important thing that's happening is the stimulating emission that's occurring as a result of that population inversion, which is being driven by the electric current flowing through there, and the back and forth bouncing of photons between those mirrors, building up that high intensity light that comes out, a small fraction of it, through that mirror. That's how a laser works.

Now, I mentioned that lasers can use a wide variety of media, and just to give you one example before a get to the specific example, all lasers have something in common, although the details differ. They have an energy source called the *pump*. That's what produces this population inversion—a flash lamp on the first laser, electric current in my simple helium neon laser, which is pretty much the same as the laser in here—some very high-powered lasers actually use chemical reactions that produced the products of the—it's literally the combustion process—they come out in an excited state, and then, they fall down and emit stimulated emissions. Some of the lasers proposed for missile defense, for example, work that way.

Another laser can be the stimulating source. Some called *dye lasers* use organic dyes that are tuned to almost any wavelength of light, and you hit them with another laser; this dye undergoes laser action.

There's, again, medium, the thing that is doing the stimulated emission—it could be a solid, like ruby; neodymium glass is used in high-powered lasers. Gases—helium neon, carbon dioxide lasers, there are nitrogen lasers, etc. There are liquids with these tunable organic dyes, for example, and semiconductors.

All lasers also have in common this so-called *resonator*, this cavity that also happens to be the right length to be resonant, like those waves on a string that I had earlier, at the frequency of a laser, but that's not hard to achieve. They have glass mirrors, or, if it's a solid medium, you can simply polished the ends, and sometimes, you polish the angle for reasons that have to do with the optics of the waves that reflect or refract at the interfaces. The light tends to come out automatically polarized, by the way, especially if you have an angular cut for the mirror.

One other example, a *semiconductor diode laser*—because we know how semiconductors work, from the module on semiconductors—here's a semiconductor diode laser, and it's basically just a PN junction with a carefully engineered energy gap. We send an electric current through it. You'll notice that this is in the correct direction to forward bias, this diode, with the positive side of the battery to the P, and the negative side to the N, and if the two faces of the diode at the junction between the P and the N are polished so that they act as mirrors, or coated with a thin metallic coating, so that they act as mirrors, then what would be a light emitting diode, just emitting light at the energy of that band gap in this particular semiconductor, the light bounces back and forth, stimulating emission between electron-hole pairs that are being created by the energy from the battery, so that the population inversion consists of there being electron-hole pairs that wouldn't naturally be there, and we get laser action, and out comes a laser beam. That's a diode laser.

Now, we think of lasers as being very bright, but this little laser pointer, for example, is producing light of about 1000th of a watt, or less. It's a tiny, tiny power. The only reason it looks so bright is because that power is concentrated in a very, very small space. This little helium neon laser is probably at most a few milliwatts, a few thousandths of a watt. So, the common lasers we tend to deal with are of very, very low wattage, low power. It doesn't mean that their beams aren't dangerous, because that light is very concentrated. You wouldn't want to shine any laser beam into your eye.

There are also lasers available with very, very high power. On the left here, for example, you see the nova laser at Lawrence Livermore Laboratory. The nova laser actually consists of a whole bunch of laser beams that converge on a spherical chamber, and the idea is literally to implode a tiny pellet of deuterium, or deuterium and tritium—more on that in Lecture Thirty-Four, the next lecture, "Nuclear Physics" —to try to make them undergo a fusion reaction.

There are two reasons the folks at Lawrence Livermore want to do that, one benign, and one not so benign. The benign reason is because, again, as I've said earlier, if we ever got fusion working as an energy source, the oceans would have 300 billion years' worth of fuel in them at our current energy consumption rate. That might be a good thing, it might not be, but it's pretty non-polluting energy, if we could get it going. The other reason the folks at Lawrence Livermore are doing this is because those fusion reactions that go on as that pellet implodes are miniature hydrogen bombs, and in an era when nuclear testing doesn't happen, it's a way of testing nuclear weapons.

At the right, you see the airborne laser. It's a several million watt chemical laser, mounted on a modified 747 aircraft, and it's under development for ballistic missile defense. The idea is that you literally zap the missile, and destroy it with this high-intensity laser beam; shot perhaps hundreds of miles from the upper atmosphere to the missile you're trying to zap. A lot of money is going into that. It's not clear whether that's a good idea, but there is. That's a very high-powered laser, and some of the highest-powered lasers have applications that I don't particularly want to think about, but there they are.

Lasers have actually been made in almost every wavelength. I mentioned "masers," "Microwave Amplification by Stimulated Emission Response." We have infrared lasers; the laser that reads the pits in an ordinary audio CD is an infrared laser. The laser that reads the DVD disc is just into the visible, a little bit into the red. The lasers that will read those Blu-Ray discs are semiconductor lasers, little diode lasers, in the blue.

We have lasers in the ultraviolet, although infrared and visible are the most common. People have made X-ray lasers; and gamma-ray lasers—nuclear radiation, basically, in an intense laser beam—are under development. By the way, some X-ray lasers have actually used nuclear explosion as the pump mechanism, a strange phenomenon.

People are also working on what might be called *atom lasers*, in which you get actual atoms—not photons, but atoms of matter—to all interact in that same coherent way to produce a coherent beam, and that's an atom laser.

What do we do with lasers? Well, lasers have almost unlimited uses today. We use them in communication, industry, commerce, science, and I want to now spend the rest of this lecture exploring a few of those applications. I want to start with an example.

Here's my helium neon laser, again, my visible helium neon laser that you can see through, and I've got it shining across the studio here. There's its beam, and it shining on a little detector. If I've got it aimed right, it's going right into that little light detector. That light detector is going into an amplifier, which is going to amplify what it detects. What's it going to detect?

Well, coming into the laser at the back is a signal from a CD player over here, and it happens to be playing my other course, *Einstein's Relativity in the Quantum Revolution: Modern Physics for Modern Scientists*. I should have turned it to the part on quantum physics, and I could be quiet for a while, and you could hear about that. I think we're talking about cosmology at that point, though.

Some of the electronics inside this plastic box are being used to modulate the laser beam. Its intensity is changing slightly with that audio signal it's coming in. You don't see the modulation of the beam. In a minute, I'm going to hit "play." It's much too minor for you to see, but this light detector can detect those slight differences in the intensity of the laser beam, and it's going to into this amplifier, which I will turn on in a minute. First, I will hit "play" over here.

Okay? The player is now playing. I will turn on the amplifier. A crazy physicist, talking about infinite dough. I want to show you that the information that is coming out of here that you can hear—kind of scratchy, not the best communications link—is in fact carried on that laser beam. I'm going to use my hand to interrupt the beam. Gone. Gone, because I'm blocking the laser beam, which is carrying that information.

You get the idea. We are clearly communicating by means of a light beam. That light based communication is, in fact, what goes on in all of our fiber optic systems. Our fiber optic systems usually use a little diode laser, a little semiconductor laser, at the entrance to the fiber optic, and it is turned on and off rapidly to represent the ones and zeros of digital information. The principal though, is exactly like what I have shown here. The only difference is that this happens to be analog. I'm sending in an analog audio signal. We are taking it out of the headphone jack of that CD player. It's an analog audio signal, and it is modulating the laser in a continuous way that is analogous to the intensity of the sound waves. That's why it's analog.

In fiber optic communication, it is invariably digital, because you can tell the difference between a one and a zero, you can get that without distortion, but it's the same principle. Laser beams can be used to carry information, and because the frequency of light is so high, you can cram an awful lot of information onto a laser beam.

Some other examples of laser use: I mentioned CDs and DVDs way back in the first module. Again, those use semiconductor lasers that have band gaps in the infrared for CDs, the visible for DVDs, and for the new Blu-Rays in the blue visible.

Lasers read barcodes. You have seen bar code scanning in supermarkets, libraries, and in stores of all kinds. They're reading the barcode. They are shining light that is then picked up by light detectors to read the barcode.

Lasers with outputs in the kilowatt range can cut metal, and they can cut it very precisely. Here's metal being cut with a high-intensity laser.

Laser beams can destroy biological tissue. That makes for a kind of surgical knife, kind of optical knife, and lasers are used in surgery.

We enjoy lasers in laser light shows, in which laser beams are steered either by mechanical or optical means. We use lasers for remote sensing of the environment; by using tunable lasers to hit individual atoms, and see what energy levels we are exciting, and can determine what kind of materials there are, and how much there is, in, say, a polluted hunk of air or water, or some other material.

We may be able to detect clandestine nuclear weapons production this way. One interesting thing laser beams can do is trap tiny objects. They're called *optical tweezers*, and I have a picture here of how that works. Here's a laser beam. It's more intense in the middle. It is being shone through a lens to focus. It's more intense in the middle, and it's also more intense in that focus. Here's a little particle, and here comes some light from the last intense part of the beam. It's refracted through that little particle, and so, that means the light has been bent in that direction. By Newton's third law, the light must have exerted a force in the other direction on that particle.

On the other hand, here comes some light from the more intense center of the beam. It's refracted the other way. That means the force on the particle must be in the other direction, and the net force ends up being down toward the focus. The particle, in fact, gets trapped just below the focus, and you can actually move it around. Here's a picture in which I show a cluster of three cells. They're going to stay fixed, and there's a little yeast cell, picked up by optical tweezers, and moved relative to those other three cells. There it goes. We can manipulate tiny, tiny objects.

Here's a movie, in which we see a microtubule—that's the long structure running diagonally across here. At the end of the microtubule are two little glass beads. Those little glass beads are attached to the microtubule, and they are being caught in the beam from a laser. That whole thing is dancing around your screen, vibrating, twisting, because the operator of the optical tweezers is manipulating that object. That microtubule is a tiny little structure from inside a cell that acts as a kind of conveyor belt, to move materials around in the cell.

Lasers are used for surveying. The straightness of the laser beam makes them a wonderful survey tool. You can walk to your local hardware store, or home store, and find a whole host of laser levelers, and other laser surveying instruments. Lasers can be used to cool atoms, because the Doppler effect of an atom moving just the right way sees the laser beam Doppler shifted, and it absorbs that energy, and comes to a stop. Some of the cooling techniques used to get to absolute zero involve laser cooling.

Finally, let me end with laser interferometry, a technique that use lasers for very precise measurement of distances. Here's a system where I have a beam splitter. It's going to split the laser into two directions. It's going to go out to those two mirrors, and it's going to bounce back to that beam splitter. It's going to be recombined; the two beams combine, and they make an interference pattern. If I move one of those mirrors slightly, even a fraction of the wavelength of light, that interference pattern will shift, and I will be able to observe that. I'm going to quickly demonstrate for you how that works with this laser system here.

I'm hooking up this laser -- this laser is a little bit more powerful. It's a green laser. You can see an interference pattern here. There it is on that screen, a pattern of lines: Bright, dark, bright, dark, bright, dark. That's the interference of those two beams coming together. If I push on this one mirror, with something as little as a feather, on this rigid mirror, on this steel post, the mirror is moving. A few wavelengths of light, and that's enough to change the interference pattern. If I pour a little liquid nitrogen on this, you'll be able to see the actual beams. I'm actually not going to pour the liquid nitrogen. I'm just going to pour the vapors, and you can see the beams going back and forth between those two mirrors, coming out of the original laser, coming out of the beam splitter, going to mirrors, and then going to this lens, which I used to magnify the patterns. There are the beams being produced in that interferometer.

Devices are used to measure very, very precise distances on optical surfaces. Let me end with a very audacious use of lasers, an interferometer so big that it is three million miles between some satellites that hold the mirrors. This is a device that is used to detect gravity waves. It isn't launched yet. It is going to be launched sometime around 2010, perhaps. The precision to which this three million mile distance will be measured is to about one billionth of an inch, using laser interferometry. That thing is a substantial fraction of the distance from the Earth to the Sun, and it will be a wonderful scientific instrument, if we get it off the ground.

Lecture Thirty-Four
Nuclear Matters

Scope: Nuclear physics was born around the turn of the 20th century and
matured in mid-century. Today, nuclear physics is inextricably part
of our lives in ways both obvious and subtle. In the United States,
some 20 percent of our electrical energy is from nuclear sources;
although the United States produces the greatest total amount of
nuclear energy, the percent reliance on nuclear power is much
higher in many other industrialized countries. The continuing
spread of nuclear weapons threatens the stability and safety of the
planet. At the same time, nuclear applications from medicine to
smoke detectors to airline security enhance our safety and well-
being. And nuclear technologies help advance knowledge in fields
as varied as biology and archeology.

Outline

I. A brief nuclear history.

  A. The first hints of nuclear physics came in the late 19th century,
  with Henri Becquerel's accidental discovery (1896) that uranium
  compounds fogged his photographic plates. Marie Curie followed
  with a doctoral thesis on the phenomenon, which she named
  radioactivity. Curie and her husband, Pierre, shared the 1903
  Nobel Prize in Physics for their pioneering work, and Marie Curie
  was later awarded the Nobel Prize in Chemistry for her discovery
  of two new radioactive elements.

  B. Physicists soon recognized three types of radiation from
  radioactive materials, which were eventually identified as electrons
  (beta radiation), helium nuclei (alpha radiation), and a high-energy
  form of electromagnetic waves (gamma radiation).

  C. In 1909–1911, Ernest Rutherford and his collaborators Hans
  Geiger and Ernest Marsden used alpha radiation from a radioactive
  source as a probe to explore the nature of the atom. To their
  surprise, they found that the atom is mostly empty space, with a
  tiny but massive center carrying positive electric charge. They had
  discovered the atomic nucleus.

D. By the 1930s, physicists understood that the nucleus consisted of two distinct particles: positively charged *protons* and electrically neutral *neutrons*. Because like charges repel, there must be some other force besides electricity that holds the nucleus together. That new force is the *nuclear force*. It's the balance between the nuclear and electrical forces that determines the stability of the nucleus— and explains why larger nuclei tend to be unstable and undergo radioactive decay.

    1. The number of protons in a nucleus—called the *atomic number*— determines how an atom interacts chemically and, thus, what element it belongs to. That's because a neutral atom consists of the nucleus and a number of negative electrons equal to the number of protons in the nucleus, and it is the electrons that determine the atom's chemical behavior.

    2. Atoms of the same chemical element can have different numbers of protons in their nuclei. Such different *isotopes* have identical chemical properties, but they differ in mass and sometimes in other physical properties.

E. In 1934, Irène Curie (Marie's daughter) and her husband, Fréderic Joliot-Curie, were the first to produce new radioactive elements by bombarding substances with radiation from natural radioactive substances. (This work won them the Nobel Prize.)

F. Soon physicists and chemists were using neutrons as probes of matter, producing new isotopes when the neutrons were absorbed and the bombarded nuclei subsequently underwent nuclear reactions. The Italian physicist Enrico Fermi was a pioneer in this work.

G. In 1938, the German chemists Hahn and Strassmann bombarded uranium with neutrons. To their surprise, they found radioactive isotopes of much lighter elements among the products of their experiments. The Austrian-born physicist Lise Meitner, who had fled to Sweden to escape Hitler, soon realized what had happen: Uranium nuclei had split, with an enormous release of energy. Meitner, publishing with her nephew Otto Frisch, coined the term *nuclear fission* to describe this new phenomenon.

H. Only the isotope uranium-235, which constitutes just under 1 percent of natural uranium, readily undergoes fission. But when U-235 fissions, it releases several neutrons, typically two or three.

Each of these can cause additional fission in nearby uranium nuclei. The result is a *nuclear chain reaction.*

I. In the late 1930s, the military implications of nuclear fission and chain reactions were obvious. The rush was on to harness the new energy source. (There's a lot of history crammed into that sentence!) By 1942, a group led by Enrico Fermi had established the first controlled chain reaction, in a reactor built under the stands of the University of Chicago's stadium. Three years later, the first nuclear weapons were detonated, and the nuclear age had begun.

II. Nuclear fission provides one of two basic ways to extract energy from the nucleus. Fission works because the energy associated with a pair of middleweight nuclei is less than that of a large nucleus, such as uranium; thus, energy can be released by splitting large nuclei. But energy can also be released by joining light nuclei in the process called *fusion.* The famous *curve of binding energy* expresses both these possibilities and is at the essence of the possibilities and dangers we face in the nuclear age. The energy released in these nuclear reactions is some 10 million times that released in such chemical reactions as burning coal or metabolizing food.

A. To get sustained energy release from fission, we need to ensure that at least one of the neutrons released in each fission event triggers an additional fission. Neutrons can be lost by escaping from the material or being absorbed by materials other than uranium-235. For this reason, fission requires a certain *critical mass* of uranium.

   1. The explosive, rapidly multiplying fission of a nuclear weapon requires uranium highly enriched in the isotope U-235. That's why enrichment technology is crucial in the effort to slow nuclear weapons proliferation. A grapefruit-size sphere of U-235 is enough to make a weapon that can destroy a city.

   2. For controlled fission, we want exactly one neutron from each fission event to cause another fission—so we have to lose some of the two or three neutrons released in each fission event. Fuel for a typical fission reactor is only slightly enriched in U-235; as a result, many neutrons hit U-238 (where they make some of the byproducts of fission, especially plutonium). Others are lost to special neutron-absorbing control rods that can be moved in and out of the uranium to control the reaction rate. And because fission occurs more readily with slower neutrons, most reactors include a *moderator*, which slows down the neutrons.

3. There are a number of reactor designs in use around the world, employing different fuel-enrichment levels, different moderators, and different coolants to carry off the heat generated by fission. Each has engineering, economic, and safety advantages and disadvantages. In the United States, nearly all power reactors use ordinary water as the moderator and coolant simultaneously. In some, the water boils directly in the reactor to produce steam that turns a turbine; in others, pressurized water transfers its heat to a secondary water/steam system.

B. Nuclear fusion is much more difficult to achieve, because the electrical repulsion of two positively charged nuclei makes it difficult to get them together. There's no chain reaction involved; just individual fusion events that release energy.

1. Fusion is what powers the Sun and, hence, nearly all life on Earth. The Sun's immense gravity is able to contain the hot, fusing hydrogen at its core. The fusion process in stars ultimately creates the elements we're made of (more on this in the last lecture).

2. Uncontrolled fusion powers thermonuclear or "hydrogen" bombs, the reaction initiated by a fission explosion and a sophisticated focusing of the fission energy on hydrogen fuel.

3. We haven't yet achieved controlled fusion on Earth, although scientists are inching toward that goal. Efforts involve either containing hot (100-million-degree!) gas with magnetic fields or using intense laser beams to produce fusion temperatures and densities in tiny fuel pellets. If fusion succeeds, we will have enough deuterium fuel in the oceans to last 300 billion years! That's 20 times longer than the Sun is going to continue shining. (Deuterium comprises about 1 in every 6,000 hydrogen atoms.)

III. Radiation is a big worry with nuclear technologies. Radiation is harmful, but it's also useful in many ways.

A. Nuclei with too many protons or too many neutrons are unstable and decay by emitting high-energy particles. That's the radiation. More massive nuclei need more neutrons to hold them together; thus, when uranium fissions, the resulting nuclei have too many neutrons and are highly radioactive. Uranium itself is only mildly radioactive.

B. Radioactive nuclei are characterized by their *half-life*, or the time it takes half the nuclei in a given sample to decay. Half-lives range from fractions of seconds to billions of years.

C. Radiation is harmful to biological systems because it damages DNA and other biomolecules. This can cause cell death, genetic mutations, or cancer. About 80 percent of the radiation we receive is from natural sources, including 11 percent from radioactive potassium in our own bodies. The remaining 20 percent is largely from medical procedures; nuclear power is responsible for less than 1 percent.

D. Nuclear radiation has a great many practical applications that affect our lives. Here are just a few:
   1. Radiocarbon dating has vastly expanded our knowledge of the past, providing accurate ages for once-living objects up to about 50,000 years old. (Other radioisotope dating methods go back through the billions of years of geologic time.) Cosmic rays interact with Earth's atmosphere to produce the radioactive isotope carbon-14, with a 5,730-year half-life. The radioactive carbon forms carbon dioxide, which is taken up by plants, then animals. As a result, living things maintain a steady level of C-14. When they die, uptake ceases, and the level drops as the C-14 decays. Measuring C-14 levels in a long-dead sample of wood, bone, or other formerly living material determines its age.
   2. Smoke detectors save thousands of lives each year in the United States. The simplest, most common, most effective, and most economical smoke detector is the ionization detector. This type of detector contains a tiny quantity of the radioactive isotope americium-241 (half-life 432 years), which emits alpha particles (helium nuclei) as it undergoes radioactive decay. The alpha particles collide with air molecules, giving them an electric charge and making the air an electrical conductor. An electric current is passed through

the air, and the detector monitors this current. If smoke particles enter the detector, they absorb the alpha particles and reduce the current, which triggers the alarm.

3. Americium has 95 protons in its nucleus and, like other elements with more protons than uranium's 92, does not occur in nature. Americium is formed in nuclear reactors by a sequence of nuclear reactions that begin when uranium-238 captures a neutron. Our source of this valuable substance is, thus, the waste from nuclear power plants.

4. Speaking of fire safety, you don't need to worry about the exit signs in a public building going out in the event of a power failure. Some of these signs are powered by the radioactive hydrogen isotope tritium (hydrogen-3; half-life 12.3 years). They're similar to ordinary fluorescent lamps, in which electrons strike a phosphor coating to produce light, except that the electrons come not from electric current but from the radioactive decay of tritium.

5. Long used for chemical analysis, the process of *neutron activation* exposes unknown materials to neutrons, which are absorbed and make the materials radioactive with short half-lives. The characteristic energies of the subsequent radioactive decay provide unique "fingerprints," giving the elemental composition of the material. Neutron activation is under development as one of the most promising and reliable methods for scanning airline luggage to detect explosives and other dangerous substances.

Suggested Reading:

Paul Hewitt, *Conceptual Physics*, chapters 33–34.

Richard Wolfson, *Nuclear Choices: A Citizen's Guide to Nuclear Technology*.

Going Deeper:

Richard Garwin and Georges Charpak, *Megawatts and Megatons: A Turning Point in the Nuclear Age?*

Richard Wolfson and Jay M. Pasachoff, *Physics for Scientists and Engineers*, chapters 43–44.

Questions to Consider:

1. In the outline above, while discussing the history of the nuclear age around the start of World War II, I noted, "There's a lot of history crammed into that sentence!" Explore that history, and in particular, determine what roles Albert Einstein played (and what he didn't do) to advance the nuclear age.

2. A typical coal-burning power plant is fueled many times a week with 110-car trainloads of coal. A typical nuclear plant of the same power-generating capacity typically gets a couple of truckloads of uranium once every year or so. Why the difference? What number in the outline above speaks to this difference?

Lecture Thirty-Four—Transcript
Nuclear Matters

Welcome to Lecture Thirty-Four, "Nuclear Matters." We human beings have been aware of nuclear matter since around the turn of the 20th century, but nuclear roles in our lives go back much further than that, in the lives of our species, our planet, and ourselves. Four point five billion years ago, the Sun formed, and soon, its nuclear reactions turned on in its core. Since then, those nuclear reactions have been supplying the sunlight energy that keeps Earth's systems, and life on Earth, going.

Even earlier than that, nuclear reactions in long dead stars cooked up the chemical elements that we are made of. That will be the subject of Lecture Thirty-Six, but through the 20th century, the role of nuclear physics became more and more prominent in our lives, both in ways that are obvious, and subtle. For example, in the United States, we get roughly 20% of our electrical energy from nuclear sources. This picture I have on the screen shows you the percentage of electricity supplied by nuclear plants in the top 10 nuclear states in the United States.

My own state of Vermont is the highest. We have somewhere between 70 and 80% of the electricity generated in our state from our single nuclear power plant. By the way, these numbers are not the percent of nuclear electricity consumed in the state. Vermont is actually a net exporter of nuclear energy, and so, about half of our electricity comes from nuclear sources, and the same argument might apply to some other states. However, you can see that there's a substantial fraction of nuclear energy being generated in quite a number of states, and if you look around the world, there are countries—with France the leader, with somewhere between 70 to 80% of its electricity—being supplied by nuclear power plants. The US is down off the bottom of this chart, at about 20%, but many European countries and some Asian countries get a substantially larger fraction of their energy from nuclear sources.

There are many other ways in which nuclear matters impact our lives. The threat of nuclear annihilation still hangs over us, decades after the end of the Cold War. Nuclear technologies are also used in a variety of benevolent ways. They are used in medicine. There are used in screening airline baggage for explosives. There are used in simple things like smoke detectors. There are used in a wide variety of medical applications. Nuclear matters are very definitely a part, today, of the physics in our lives.

In this lecture, I would like to give you a little feel for how nuclear technology works, how nuclear physics works, a brief history of nuclear physics, and then, a number of applications.

First, the history: The first hint that there was nuclear physics, that anything nuclear was going on, occurred in 1896, when Frenchman Henri Becquerel accidentally left some uranium compounds in a drawer where he also kept film. He found that the film was fogged, and he reasoned that some sort of contamination must have been coming from the uranium.

A young physicist, Marie Curie, followed with a doctoral thesis on this subject, and she and her husband Pierre shared the 1903 Nobel Prize in physics for their pioneering work on radioactivity. Marie Curie later shared the Nobel Prize in chemistry for discovering several new radioactive elements. In the early days of nuclear physics, it wasn't clear whether this was physics or chemistry, and physicists and chemists both contributed to the efforts.

Scientists studying the materials soon realized that there were three kinds of emanations that came out of these materials. One, they called beta emission, and that later turned out to be just high-energy electrons. One they called *alpha emission*, and that turned out to be helium nuclei—more on that in a minute—also coming out with high energy. The third, called *gamma radiation*, was actually a form of very high-energy electromagnetic radiation, even over shorter wavelength than X-rays.

In the early 1900s, scientists began using these high-energy particles as probes of the structure of matter, something that has continued to this day—although nowadays, we accelerate the particles in accelerators, rather than using radioactive sources. Particularly, in 1909 to 1911, Ernest Rutherford and his collaborators, Ernest Marsden and Hans Geiger—the latter of Geiger counter fame—did a series of experiments where they took a source of these alpha particles, helium nuclei, and bombarded a gold foil with them. To their great surprise, they had expected all of the particles to go through. Most of them went through, but a few bounced back, almost in the direction they had come from. It was as if you had spread out a role of paper towels, and let go with machine-gun fire. Most of the bullets would go through, once in awhile, one came bouncing back, as if it had hit something extremely massive and hard in there. Rutherford and his colleagues had therefore discovered the atomic nucleus.

Before that time, the nucleus was thought to be something like the picture on the left here. This is the model of Thompson and Kelvin, around 1900, who imagined the atom as a kind of spherical blob of pudding-like material, the plum pudding model, positively charged, with an equally negative charge spread throughout in the form of the electrons. By 1911, Rutherford and his colleagues had discovered that the nucleus was more like solar system, mostly empty space, a very tiny, but very massive nucleus at the center, and the electrons in some kind of orbital system around them.

Well, we want to focus in more on the nucleus. We've already talked about atoms and electrons. We are going to look in on the nucleus. By the 1930s, physicists understood that the nucleus was composed of two different kinds of particles: Photons, positively charged particles, and neutrons, electrically neutral particles that are very similar to protons in other ways. Each of those particles, by the way, is about 2,000 times more massive than the electron is.

Now, protons repel each other by the electric force, so a nucleus couldn't stick together unless was some other force there, and that other force is the so-called *nuclear* force, and if you will get my modern physics lectures, you'll see that that is a residue of something more deep, called the *color force*, that binds quarks together, but we won't go there. We will simply point out that there's a very strong but very short-range force that binds nucleons—protons to protons, neutrons to neutrons—together. There is still the repulsive force of the protons, and if you put too many protons in a nucleus, it will be unstable, and will tend to fly apart. That's why all elements heavier than uranium are, in fact, unstable. They are naturally radioactive. They don't occur in stable forms.

The number of protons in the nucleus, the amount of positive charge, determines how many electrons get attracted around it, and determines the element's chemistry, determines its chemical nature, its interaction to make molecules. That's interesting, but it's not what interests nuclear physicists. That number, by the way, is called the *atomic number*. The important thing for physicists is that different atoms of the same element—that is, the same atomic number—can have different numbers of neutrons. Atoms of the same element, with different numbers of neutrons, and therefore different weights, are called *isotopes*. I'm going to give you some examples of isotopes here.

I've got a picture, first, showing a proton. I'm going to symbolize that as a red sphere, positively charged. A neutron will be a gray sphere, and here are some examples of isotopes. Hydrogen, the simplest hydrogen, its nucleus consists of a single proton. It's called "*hydrogen-1*." That's the common isotope of hydrogen, but there's also deuterium. It's given its own name. It's also called "*hydrogen-2*." Deuterium consists of a proton and a neutron. Because it still has one proton, it gets one electron around it, and its properties chemically are the same as ordinary hydrogen, but it's nuclear properties are quite different, and its mass is twice as great.

Tritium, *hydrogen-3*, is a radioactive version of hydrogen. It has a half-life—more on that later—of about 12 years, so it doesn't exist in nature unless it is formed by some process.

Here are the isotopes of oxygen; a common form of oxygen is oxygen-16, but there's another stable isotope called oxygen-18; again, they are chemically identical, but water made with oxygen-18 and water made with oxygen-16 evaporate at different rates because of their different masses. It's that effect, and measuring the relative ratio of these two isotopes, that allows climatologists to look in ancient ice cores, and figure out what the temperature was. I'm not going to go into details, but I showed you some of the results of that date back in the lecture on climate change.

Now, one of the more massive nuclei is uranium, and uranium comes in a number of isotopes, but two are particularly important. Uranium-235, which constitutes less than 1% of natural uranium, has 143 neutrons, and 92 protons. Uranium-238 has 92 protons as well, but 146 neutrons, so it has a total number of particles, its mass, basically, of 238. Again, these are chemically similar, but their nuclear properties are very, very different.

Now, with the discovery of the neutron in the 1930s, physicists began using these neutrons as probes of the nucleus, and with good reason. Because the neutron is neutral, it doesn't get repelled by the positive charge of the nucleus, and easily penetrates them. One of the leading pioneers in this was Enrico Fermi, a name that may be familiar to you. I have shown you pictures of Fermi lab before, where the giant accelerator is, for example, named after Enrico Fermi. He was an Italian physicist who came to the United States about the time of World War II.

In 1938, two German chemists, Otto Hahn and Fritz Strassmann, had begun bombarding uranium with neutrons. To their great surprise, they found that sometimes, instead of just making a slightly more massive nucleus as an extra neutron was absorbed, or maybe making a slightly less massive nucleus as an alpha particle came out, sometimes they found nuclei that were about half the mass of uranium, and they were very puzzled about this.

A former colleague of theirs, Lise Meitner, who had by that time fled Germany to escape Hitler and was in Sweden, was the one who came up with the explanation. By the way, it's one of the great travesties of scientific history. Meitner is not always recognized for this discovery, and she also did not share in the Nobel Prize that they were awarded, eventually, for it. She certainly should have been. Much later, she was honored by having "Element 109," an artificially produced element, named after her. That element is now called *meitnerium*.

Anyway, Meitner was walking in the Swedish countryside with her nephew, Otto Frisch, who was also a physicist, and she drew a picture, reasoning what was happening, how the uranium was splitting. She coined the term *fission*, and they published a paper in which the term "fission" was first used.

In nuclear fission, the uranium-235 nucleus is hit with a neutron. In comes a neutron. The combined nucleus, now with the extra neutron—it's uranium-236—is every stable, and it begins vibrating with the energy that has been imparted to it. That vibration makes it an elongated structure, and then, the positive electric charge associated with the protons is far apart, and electrical force, although it falls off with distance, does so less rapidly than the nuclear force, and so, the electrical force becomes important, and this nucleus vibrates a while, but eventually, it forms this dumbbell shape, and then the two halves go apart. They aren't necessarily equal halves, and a whole range of what are kind of middleweight elements is formed in this process of nuclear fission.

Now, remember that this discovery was made in the late 1930s, on the eve of World War II. One could do a whole course in the history of nuclear physics, especially around World War II, so in this one sentence I'm going to utter is crammed a lot of history. The sentence is this: The military implications of this discovery were obvious, and I think they were also very ominous. That's what started the Manhattan Project. I can't go into the history of that—Einstein's role, Wegner's role, various other roles—but eventually, the Manhattan Project was started. By 1942, Fermi had led a group building a controlled nuclear fission reactor under the Stands of the University of Chicago. By 1945, the first nuclear weapons were detonated.

Both of those nuclear weapons involved chain reactions, and to understand what a chain reaction is, let's look at something else that comes out in this process of nuclear fission. In addition to the uranium fissioning in half, splitting in half, typically, one, two, or three loose neutrons are released, either immediately, or after a very short time, in this process.

Well, what can happen then is if a single neutron comes in and hits a uranium-235 and it fissions, outcome those additional neutrons, and they hit additional uranium-235 nuclei, and cause additional fissioning. Pretty soon, that whole thing can go up in a giant release of energy. By the way, it is only uranium-235 that readily fissions, and that's the reason why we worry a lot about the so-called "enrichment" of uranium. Uranium-235 is only 0.7% of natural uranium, so that fissioning process doesn't really occur easily in a mass of natural uranium, but if you enrich it to mostly uranium-235, you can get an explosive chain reaction, if you have a *critical mass*, a big enough amount that the neutrons can't escape. Unfortunately, a critical mass of uranium weighs about 50 pounds, and is about the size of a grapefruit, and if a crude bomb like the one used at Hiroshima can destroy an entire city, that's one of the reasons you read all the time in the news about people worrying about uranium enrichment technology. That's a key to nuclear weapons.

For controlled fission, we want something else to happen. We want exactly one neutron coming out of each fissioning uranium to cause another fission chain reaction. So for controlled fission, we want to have materials in with the uranium that will absorb neutrons and prevent fissioning from occurring for most neutrons, so that the reaction won't run out of control. Typically, we use fairly unenriched uranium. In the United States, we enrich uranium to about 3%, U-235, for our fission reactors. Canada has a natural uranium kind of reactor called CANDU that works on different principles, and doesn't need any enrichment, and so on. We also put in neutron-absorbing materials that can be moved in and out of the reactor, to controlled rate of the nuclear chain reaction.

There are a lot of different reactor designs in the world, and I don't think it's fair to say that any of them are superior. I mentioned the Canadian reactor. In the United States, we tend use reactors that are called *light water* reactors, because they use ordinary water. Something else a nuclear fission reactor needs that a bomb doesn't need is something called a *moderator*. It turns out that these nuclear reactions occur better with neutrons that are moving relatively slowly. That really isn't an issue in a bomb, where you have a lot of stuff all together in one place, but in a reactor, unless you could slow down the neutrons, they wouldn't be effective enough at causing fission.

There is therefore something called a "moderator." You might think that it moderates the reaction, but it doesn't. It actually makes it go better, because it slows down the neutrons, and water makes a very good moderator, for reason I described in an earlier module, when I showed two objects of equal mass colliding, or I showed my Newton's cradle thing. That is because a neutron and a proton have the same mass, and when a neutron strikes a proton in water, he transfers almost all of its energy, and slows down significantly.

US reactors, then, use light water. US reactors are basically an outgrowth of submarine based reactors. The Candu reactor that I mentioned, that the Canadians make, uses heavy water as its moderator, and that allows it to use natural uranium without enriching it. That's because the heavy water doesn't absorb neutrons as easily. Heavy water is deuterium-based water. It's kind of rare, though. You have to remove it from ordinary water at some cost, expense, and trouble.

The Russian reactors tended to be graphite-moderated. They had a solid moderator. The reactor at Chernobyl, which caught chemical fire, was a graphite reactor. It's not clear which of these reactors is best. They all have engineering, safety, efficiency, proliferation, and other trade-offs. In a US reactor, a US light-water reactor, the nuclear fission reactions are occurring in rods containing uranium fuel. These rods are stationary in a vat of water—a big pressure chamber of water that is either boiling or under pressure—water reactors versus pressurized water reactors.

Between the uranium fuel rods are so-called *control rods*, and the control rods can move up out of the way, absorbing fewer neutrons, making the reactor go a little faster, or down, slowing down the reactor. If they are pushed all the way down, they will make the reactor stop completely.

Well, that's one way of making energy from nuclear reactions. Another way of making energy from nuclear reactions involves fusing light elements together. That's the process of *fusion*. A very famous graph—the title of a book by John McPhee, *The Curve of Binding Energy*—describes these two processes. What it says it is, if you start with very heavy elements like uranium and split them, you can release energy. If you start with very light elements like hydrogen, and fuse them together, you can also release energy. That's the process that powers the stars. We have not yet made fusion worked successfully here on Earth, but if we did, we would have 300 billion years' worth of energy in the oceans, because about one of every 6500 hydrogen nuclei in ocean water is, in fact, deuterium, and we could join deuterium to make energy and produce helium in the process.

By the way, you'll notice that iron is the element that when you make it, you release binding energy, the most energy, and you can't do anything with iron. You can't extract any more energy from it.

How would a fusion reaction work? Well, it would look something like this. Here are a bunch of deuterium nuclei running around together in a very hot system. We have to contain them at about one hundred million degrees. We haven't totally figured out how to do that. I showed you, in earlier lectures, two possible methods. One was the magnetic confinement of the ITER experimental fusion reactor. Another was the nova laser system I showed you in the previous lecture, which achieves this by blasting a tiny pellet with very high-energy laser beams. Occasionally though, if you get to enough temperature, these rapidly moving nuclei overcome the electrical repulsion of their being positive, form a helium nucleus, and emit a large amount of energy in the process.

That is how nuclear fusion works, but we simply aren't there yet in terms of producing nuclear fusion as a reliable energy source, something we can control. Unfortunately, we have learned to make it go off with an explosive reaction, and that's one so-called *hydrogen bomb*, or thermal nuclear bomb, is about.

Now, talk about nuclear things, and everybody worries about radiation. Radiation is indeed something we should worry about, but it is also something that can be useful. The most important characterization of radiation is the half-life of a radioactive nucleus. The half-life is the time it takes half of all the nuclei to decay. If I start out with 100 nuclei, for example, then one half-life later, whatever the half-life is, there will be 50 of them left, approximately. They decay at random by quantum mechanical processes that are intrinsically random, but roughly 50 will be left. After two half-lives, roughly 25 will be left. After three, half of 25, or 12.5, then 6.25, I guess, and down it goes. By the time a few half-lives have gone by, the total original amount of radioactive material is largely gone. It has decayed by emitting particles into some other kind of substance.

Some typical half-lives of some radioactive materials, some of which are quite useful: Uranium-238; that's still around because it has a half-life of 4.5 billion years. Uranium-235 is only 700 million years. A billion years from now, there won't be nearly as much uranium-235 around, but there will be almost as much 238. By the way, billions of years ago, there was enough 235 around that natural fission reactors actually occurred on Earth, at least one place in Africa, and ran for thousands years, because there was enough concentration of U-235.

Strontium-90 is a fission byproduct. It lodges in bones. It was a big worry during the nuclear fallout times of the nuclear bomb tests in the atmosphere, and it's also a worry in nuclear power plant accidents, because it behaves like calcium and lodges in your bones. Plutonium-239 is produced in nuclear reactors when uranium-238 absorbs a neutron, and turns through a series of reactions into plutonium-239. Plutonium-239 is even more dangerous as a nuclear weapons fuel than uranium-235, and that's the reason we also don't want a lot of nuclear reactors producing plutonium-239, and then have countries extracting it, and using it to make nuclear weapons. Both of those processes are, though, for better or worse, going on around the world.

Carbon-14, with a 5,700-year half-life, roughly, is formed by cosmic rays in the atmosphere. It's used for radiocarbon dating. I will show you in a minute how that works. Oxygen-15 is a very safe isotope. You have to make it right in a hospital. It's used for medical studies, PET scans, particularly. With its two-minute half-life, it's gone pretty soon.

Radon-222 lives only 3.8 days, but it's part of a sequence of products produced by the decay of natural uranium, and there's plenty of natural uranium in rocks. This is a gas, and it seeps into our basements. It's actually a major source of radiation for human beings. In fact, if you were to ask how much radiation you receive, and where it comes from, the answer is probably not what you might think. About half of your radiation dose on average—and this varies a lot with where you live, what the geology is, what the altitude is, and so on—about half is from radon. The local rocks and soil immediately provide about 8%, with cosmic rays providing about another 8%, although if you live in Denver, that's about doubled, because there's less atmosphere above you, and the cosmic rays are less attenuated.

Your body is a major source of radiation to you. Your body has radioactive substances, particularly naturally radioactive potassium that produces about 11% of your radiation. Medical products and techniques—X-rays in particular—give you about 15%. Some consumer products are radioactive, and you get about 3%. "Other" is about 1%. That, by the way, includes radiation from nuclear power plants and operations of the nuclear power industry. It's pretty negligible in terms of the overall effect.

Now, radiation has a whole lot of practical uses. I want to just give you some of them. I'm not an advocate of nuclear radiation. I'm kind of neutral on the whole issue, but I want to give you a sense here that things nuclear are not all bad. They aren't all good, either.

Here's how radiocarbon dating works. There are these cosmic rays coming from the Sun, partly, but mostly from distant parts of the galaxy, and maybe even beyond the galaxy. They consist of very high-energy particles. They come barreling into the solar system, and they bombard the Earth's atmosphere. They come in with such energy that they induce nuclear reactions in materials in the upper atmosphere, and one nuclear reaction forms a product called carbon-14.

Carbon-14 is chemically just like carbon. Normal carbon is carbon-12. It has six neutrons, and six protons. Carbon-14 has six protons, because it's carbon. That's what makes it carbon, so it has six electrons. That's what determines its chemistry. However, it has eight neutrons, so that it's a little more massive, and it happens to be radioactive. It's unstable. It's got too many neutrons, and so it decays. It decays with this half-life of 5,730 years.

Here's what happens. Here we have this mammoth, some time ago, walking along, and the carbon-14 was formed in the atmosphere, and became carbon dioxide, interacting with oxygen, and growing plants took up the carbon dioxide, so then the grass became slightly radioactive with this carbon-14. Of course, most of the carbon was not carbon-14, but a little bit of it was. The mammoth ate the grass, and the mammoth got carbon-14, and became radioactive. Well, the mammoth was a bit radioactive, and that's is what that is designed to show.

The mammoth died. The mammoth stopped eating grass, stopped taking up radioactive carbon, and the radioactive carbon, now in the mammoth, decayed. While the mammoth was alive, although the radioactive carbon was the king, the mammoth was always eating more radioactive carbon, and so, it stayed approximately at the same level of radioactivity. However, once it died, that uptake stopped, and the mammoth's radiation began to decay over time. There it was, later, less radioactive.

After a while, the mammoth became just a pile of bones, still just very slightly radioactive, depending on how long they lay there. Along came some archaeologists. They dug up these bones. Now, I want to point out that the archaeologists were radioactive, because they were eating plants, even if they weren't vegetarian. They were eating meat, and the meat was coming from plants, and that had been taking up radioactive carbon.

By the way, there are a lot of subtleties here. The amount formed depends on what the Sun is doing, because the sun's activity keeps cosmic rays out of the solar system, so to get these dates exact, we have to be pretty careful. The big idea works pretty well, though.

There were anthropologists or archaeologists, and they were kind of radioactive themselves. They dug up the mammoth bones, went to the laboratory, and compared the radioactivity in these old bones with that in a fresh bone, perhaps for an animal that had just been killed. They actually didn't even need to do that. They simply knew what radiation levels were in contemporary organic materials. By the way, this works for things like charcoal from a fire—because that's wood—as well as from an animal's material.

In this particular case, then, I indicated the amount of radiation by the number of arrows you see, and in this particular case, there were half as many arrows in the old bone, meaning that they were half the radiation level. That implies that one half-life had gone by, and that this bone was, therefore, one half-life old, or 5,730 years.

Here's a bone that's older. It's only got one quarter as much radiation, two half-lives. That's 11,000-something years. Here's a bone that's only one-eighth as radioactive. It takes three half-lives to get you down to that level, so that's about 17,000 years old. Obviously, you could do the math to get any numbers in between.

Radiocarbon dating works quite effectively back 50,000 years or so, and there are other radioisotopes that we can use to date things further back. That's a very, very effective technique, and has given us a real sense of our place in history, because we can date objects that go back a few hundred to a few tens of thousands of years, very nicely, with radiocarbon dating.

I will give you some other examples of where radiation is important in your lives. Smoke detectors save thousands of lives every year in the United States. The most common form of smoke detector, a little thing you put on your ceiling, has a little battery and test button, and when smoke gets into it, it beeps. How does that work?

Smoke detectors contain a tiny quantity of a radioisotope called americium-241. Americium is a substance that is not produced naturally in nature. It has too short a half-life. It's heavier than uranium. It has atomic number 95. Uranium is 92. In fact, it's a byproduct of nuclear reactors, so the americium in your smoke detector ultimately came from nuclear waste. There's a very tiny bit of it there. It produces a very tiny amount of radioactivity, and is some of that 3% of the radiation dose that you get from consumer products. It does come from smoke detectors. That radiation is well worth what smoke detectors do, and they save thousands of lives.

How do they work? Well, the americium emits alpha particles, high-energy helium nuclei. They can knock electrons off atoms in the air, or they can collide with their molecules, and stick to them. Either way, the air molecules get an electric charge. The air becomes a conductor of electricity. The battery in your smoke detector—that you are supposed to change once a year—is putting some electric current through that air. If smoke gets in there, it interferes with the alpha particles, and cuts down on the conductivity, and makes the electrical resistance higher. The current goes down, the smoke detector detects that, and triggers the alarm. That's a very simple example of how radiation is helping you.

By the way, if you are worrying about fire safety, you don't have to worry about the power going off, and causing you not to know where the exit sign is in the case of a fire. In some buildings, the exit signs are powered not by electricity, but by tritium, which again emits high-energy radiation. These detectors are basically like fluorescent lamps, except instead of the energy coming from an electric current, it's coming from these high-energy particles emitted, in this case, by radioactive tritium. Some exit signs, then, like the one shown here, are actually powered by radioactive tritium.

Now, radiation has also been used for a long time in chemical analysis. I mentioned that in the 1930s, scientists began bombarding materials with neutrons, because neutrons are neutral and easily get absorbed by the nucleus; they create new materials. When a particular, ordinary, everyday nucleus absorbs a neutron, it turns into an isotope of itself or something else. It may undergo a few nuclear reactions, and turn into something else, or it may just absorb the neutron, and turn into a heavier isotope of itself.

However, that isotope may be radioactive. It gives off characteristic radiation, and if we examine that radiation, and measure its energy, we can tell and measure what kinds of radiation are emitted and we can tell what kind of material we have. The process of neutron activation then consists of bombarding material with neutrons, seeing the radiation that comes out. We make the materials temporarily radioactive—by the way, it is one of the few processes involving radiation that does that. It makes the material temporarily radioactive, but with relatively short half-lives, we measure the distribution of emission, and we can tell what and how much material is in there.

One of the more promising methods for screening airline luggage—not your hand luggage, but the big luggage that's going into the hold of the aircraft— for bombs, which tend to have particular concentrations of certain materials like nitrogen in them, is to neutron activate the luggage—expose it to an influx of neutrons; they don't have to be high-energy neutrons, because they can just get into the nucleus, because they aren't repelled by the nucleus' electric charge—expose that to neutrons, and put the baggage against radiation detectors to determine what kind of material is in there. That may be a technique that ends up being used soon to detect explosives and other dangerous materials in airline luggage.

That's a quick survey of nuclear matters in your life. Nuclear physics is certainly a part of the physics in your life, in ways that both threaten you— as in nuclear weapons—ways that may be benign, and in some ways that are downright helpful.

Lecture Thirty-Five
Physics in Your Body

Scope: Although physics is ultimately behind all aspects of the human body, it's often more insightful to describe the body's workings in terms of biochemistry, physiology, neurobiology, and related fields. However, some aspects of the body are best described in terms of physics. Electric circuit concepts play a major role in the nervous system and the heart. Simple ideas from mechanics show how the muscles and skeleton act to permit movement. Principles of fluid dynamics govern blood flow. And the passage of nutrients into our cells, and the excretion of waste from the cells, is often described in terms of an electric circuit. Perhaps even more significantly, physics and physics-derived technologies provide unprecedented and minimally invasive techniques for imaging the interiors of our bodies. Finally, physics-based techniques allow the precise delivery of energy used in curing diseases, especially cancer.

Outline

I. Principles of physics govern the workings of our bodies. Although biology and related fields provide a more insightful overview, many individual aspects of our bodies can be understood in terms of physics.

   A. Concepts of Newtonian mechanics describe our bodies' overall motion or lack of motion. Examples include balance and center of gravity, tension and compression forces in muscles, and principles we instinctively apply to minimize forces on our bodies and avoid injury. The forces exerted on and by our bodies can be surprisingly large.

   B. Our nerves provide the communications link between body and brain, brain and body. Specialized cells called *neurons* actually transmit the signals. The neurons have extensions called *axons*, up to several feet long, that act very much like electrical cables. Our understanding of axons comes from analysis of electric currents and voltages, both by measurement and by modeling the axon as an electric circuit. Similar circuit analogies apply to cells in general and describe the flow of nutrients and waste in and out of the cell in terms of electric currents.

C. The heart is particularly amenable to electrical analysis.

    1. The heart muscle consists of elongated cells that, in their resting state, have positive electric charge on the outside and negative charge on the inside. Because charge is distributed evenly all around, the cell is not electrically charged, and there's no uneven distribution of electric charge (recall Lecture Thirteen).

    2. When the heart beats, the cells temporarily lose their electric charge. They do so in a "wave" that sweeps from the top of the heart to the bottom. While they're losing charge, the cells have a slightly more negative charge on the upper end and slightly more positive on the lower end. This pattern results in electrical voltage differences that can be measured on the skin. Electrocardiography machines measure this voltage as it varies with time, and physicians use the result to diagnose heart conditions.

    3. Special cells comprise a *pacemaker* that sets the rhythm of the heartbeat. If this natural pacemaker fails, an artificial pacemaker is implanted in the chest; it works by stimulating the heart with an electrical signal. Sometimes the heart goes into a rapid, shallow, and fatally ineffectual beating called *fibrillation*. Defibrillators delivering a heavy dose of electric current can restart the normal rhythm—another example of the heart's electrical nature.

D. Principles of pressure and fluid flow determine the behavior of our blood. The two numbers associated with blood pressure are the maximum and minimum pressures (in millimeters of mercury that pressure can support) that the beating heart transmits to the blood.

    1. In the common disease arteriosclerosis, fatty plaque builds up in the arteries. Plaque can break loose and clog a smaller artery, resulting in stroke or heart attack.

    2. Physics shows another danger: Plaque narrows the artery, meaning that blood must flow faster past the plaque than it does in the unobstructed part of the artery. But by Bernoulli's principle (Lecture Ten, on airplane flight), that means the pressure in the artery drops. If this drop is too great, the artery can collapse and cut off blood flow.

II. Physics-based technology provides a myriad of new ways to image the body's interior, many of them noninvasive.

   A. X-rays, a form of electromagnetic radiation (Lecture Eighteen), penetrate soft tissue but not bone. Captured on film, x-rays thus provide diagnostic information about teeth, bone, or foreign objects in the body. Use of x-ray–opaque substances, such as barium, allows imaging of soft tissue, such as the intestine or stomach.

   1. The German physicist Wilhelm Roentgen accidentally discovered x-rays in November of 1895. In studying the new rays, he quickly found they could form an image of the bones in his hand. By December, he showed a physical-medical society an x-ray image of his wife's hand, and a mere month later, x-rays were hailed as a medical miracle.

   2. Modern x-ray techniques can be almost as simple as Roentgen's, but more sophisticated applications have been developed. Computerized axial tomography (CT or CAT scan) takes a series of x-ray images from many points on a circle surrounding the patient; a computer then reconstructs a three-dimensional picture from the individual two-dimensional images.

   B. *Magnetic resonance imaging* (*MRI*) is a much newer and very versatile imaging technique. Unlike x-rays, there is no hazardous radiation, nor is it necessary to introduce any foreign substance into the body. MRI is noninvasive. And MRI, again in contrast to x-rays, does an excellent job of imaging soft tissue.

   1. The principle behind MRI is called *nuclear magnetic resonance* (*NMR*; the term *nuclear* was scrupulously avoided in naming the medical technique). NMR uses a very strong magnetic field (tens of thousands of times stronger than Earth's magnetic field), usually from a superconducting electromagnet (Lectures Fourteen and Fifteen). Atomic nuclei (in particular, protons or hydrogen nuclei) act as tiny magnets and tend to align with such a strong field. But their intrinsic spin (a quantum-physics property that can be thought of, roughly, as an actual rotation) causes them to precess about the magnetic field the way a top or gyroscope precesses in the presence of gravity (Lecture Thirty-Two). They can precess either aligned parallel (spin up) or antiparallel to the field (spin down). Under normal conditions, the numbers are nearly equal, but a few more are aligned with the field. However, providing energy in the form of radio waves of just the right frequency (photon energy) causes nuclei to flip antiparallel (spin down). When they flip back up, they emit a radio signal. Detecting this signal locates the precessing protons and tells about their magnetic environment. Used this way, NMR is a powerful tool for determining molecular structure.

2. Medical imaging generally uses nuclear magnetic resonance of the single protons that constitute the nuclei of hydrogen. Hydrogen is especially abundant in fat and water, both common in the body, so hydrogen nuclei are everywhere. In an MRI apparatus, the strong magnetic field is inside a cylinder that can accommodate the entire body. The signal from the flipping protons becomes a measure of the number of protons. The magnetic field varies slightly with position, and only those at just the right location can absorb radio waves of the appropriate frequency. That way, the location of the flipping protons can be determined. The strength of the signal from each position is then a measure of the density of protons or, very nearly, the density of tissue. A computer organizes the data from all regions under examination to produce images of the body's interior.

3. MRI is not only a diagnostic tool for disease but is widely used in physiological and neurophysiological studies, especially of brain function.

C. Diagnostic techniques of *nuclear medicine* use radioactive substances to image body structures and processes.

1. Radioactive versions of chemicals that concentrate in particular tissues can be used to image those tissues. Radioactive phosphate is absorbed selectively in bone; to make a bone scan, the substance is injected into the bloodstream, then makes its way into the bones. Radiation detectors outside the body then form an image of the bones. Similarly, radioactive iodine settles in the thyroid gland and can be used to diagnose thyroid problems. Radioactive thallium effectively images the brain and its blood-flow patterns.

2. A particularly sophisticated nuclear imaging technique is *positron emission tomography* (PET). Positrons are the antimatter opposite of electrons, and they're emitted in the decay of some radioactive substances (see The Teaching Company's *Einstein's Relativity and the Quantum Revolution* for more on antimatter). Positrons emitted in body tissue soon meet electrons and annihilate, sending out two gamma rays (a very high energy form of electromagnetic radiation) of precise energy and in opposite directions. Arrays of detectors pick up the gamma rays, and a computer reconstructs the points where they were emitted, forming an image of the radioactive substance. Many of the radioactive substances used in PET are short lived and, thus, very safe (some must even be made right in the hospital, using particle accelerators). PET can be used to study dynamic processes, such as breathing, blood oxygen uptake, blood flow, and even brain processes used in thinking.

III. Physics also plays a role in curing disease.

A. Lasers (Lecture Thirty-Three) replace scalpels as precision tools for surgery. One common example is laser vision correction (Lecture Four). Optical fibers can guide surgical laser beams deep into body cavities, and laser surgery is particularly effective at minimizing post-surgical bleeding. In *laser angioplasty*, a catheter tipped with a high-intensity ultraviolet laser is threaded into the arteries. Short bursts of laser radiation vaporize plaque, clearing congested arteries.

B. Strong sound waves (shock waves) can be used to break up kidney stones, allowing them to pass without resort to surgery. This can be done either with sound-emitting probes inserted near the stones or externally with sound penetrating through the skin. Some blood-clot-destroying drugs are much more effective if clots are irradiated with ultrasound while being supplied with drugs.

C. Both electric currents and ultrasound stimulate bone growth and are used to help heal broken bones in particularly difficult cases or where natural healing takes a very long time.

D. One of the most significant physics-based medical treatments is radiation therapy to kill or shrink tumors. High-energy radiation, either in the form of particles or electromagnetic energy (x-rays and gamma rays) kills cells by stripping electrons off atoms and changing the physical and chemical structure of molecules, particularly DNA. Cells may be killed outright or they may fail to divide. Radiation is now used in roughly half of all cancer cases, either to cure the disease or to enhance the chance of cure using other treatments.

1. Radiation is particularly effective against rapidly dividing cells. That's why radiation exposure is especially dangerous for children, but it's also one reason why radiation selectively kills cancer cells. Cancer cells are also less organized, and that makes them less able to repair cell damage.

2. Early radiation therapy used external beams of gamma rays from highly radioactive materials. With these beams, both position and penetration depth were difficult to control.

3. Much modern radiation therapy uses high-energy beams generated in particle accelerators that are miniature versions of the machines used in high-energy physics. Aiming beams from different directions allows energy to be concentrated at the disease site while minimizing damage to healthy tissue the beams pass through. Beams consist of either electromagnetic radiation (x-rays or gamma rays) or particles of matter. Electrons are often used because they're light and easy to accelerate, but in the past decade, proton beams have come into widespread use. The heavier protons have the advantage that they deliver most of their energy at the disease site, minimizing damage to surrounding tissue.

4. Additional ways to deliver radiation therapy include radioactive materials attached to the skin; radioactive "seeds" implanted directly into a tumor site; or radioactive solutions ingested or injected, which then make their way to a particular site.

Suggested Reading:

Louis A. Bloomfield, *How Things Work: The Physics of Everyday Life*, chapter 19, section 1.

Paul Hewitt, *Conceptual Physics*, chapter 33.

Going Deeper:

Paul Davidovits, *Physics in Biology and Medicine*.

Bettyann Kevles, *Naked to the Bone: Medical Imaging in the Twentieth Century.*

Howard Sochurek, *Medicine's New Vision.*

Steven Vogel, *Life's Devices: The Physical World of Animals and Plants.*

Questions to Consider:

1. The operation of airbags in a car is similar to the considerations in this lecture about bending one's knees when landing a jump. What's the similarity?

2. Proton beams penetrate the body more deeply than electron beams of the same energy. Why?

Lecture Thirty-Five—Transcript
Physics in Your Body

Welcome to Lecture Thirty-Five, "Physics in Your Body." The title of this whole course is *Physics in Your Life*, so by now, you recognize that physics intrudes upon, or plays a role in, your life in a great many ways, but this lecture is going to get a little bit more invasive. We're going to take physics right inside your body.

Before we do that, I should point out that physics is ultimately the fundamental science, and in principle, most scientists would argue that you could describe the workings of anything in the Universe in terms of basic physics principles. That's not a very useful, insightful, or convenient thing to do, and in fact, it's something that we can't do in general for things that are much more complicated than the simplest molecules, for example. I therefore have to point out that a full description of the body, these days at least, requires the fields of physiology, molecular biology, cell biology, biochemistry, chemistry, and so on.

Nevertheless, physics plays a significant role in the body in a number of ways. I'm just going to describe a few of them here today, in this lecture.

First of all, the body is, after all, a mechanical system. We move around, we produce energy, we raise our arms and change their potential energy, and so on and so forth. In particular, forces act on our bodies, and I want to talk a little bit about that at first.

Secondly, our bodies are partly solid—our bones; partly semi-fluid-like stuff, and partly real fluids flowing, like our blood. That branch of physics—fluid flow—plays a significant role. Perhaps less obvious than mechanics, but equally significant, is that our bodies are very much electrical entities, particularly our heart, and our nervous system. I will talk a little bit about that.

More significantly, though, I think, in today's technological world, the way physics gets into our bodies is through medical techniques, particularly techniques that allow us to image the interiors of our bodies in more or less noninvasive ways. I will describe a number of the methods currently in use, starting with simple X-rays, and moving on to much more sophisticated methods, like magnetic resonance imaging, and PET scans. I will describe how those work.

Finally, physics plays a role in the treatment of disease, and I will end with a description of those phenomena, how those technologies work.

Let me begin with simple Newtonian mechanics. The motions of the body are controlled, essentially, by the laws of Newtonian mechanics. We don't move at speeds near the speed of light, or that aren't relative to anything in our nearby environment. Relativity doesn't play a role. The laws of Newtonian mechanics fit very nicely in describing our bodies, and I want to make only one significant point here, in describing the physiology of the body. There's a lot more I could go on with—sports physiology, the physiology of respiration, and all of these things—but I just want to talk about forces acting on the body, because our bodies are subject to forces that are far larger than you might imagine. I'm going to give you two examples that show how that comes about.

Here's a bowling ball, which I've been using throughout this course for a variety of purposes. Now, I'm simply going to take the bowling ball, and I'm going to hold it like that in my arms. I would assert that the forces involved in my doing that are far larger than you might imagine. This bowling ball weighs about 15 pounds. Pounds are the English unit of force, and so you might say, "Well, the forces involved are 15 pounds," but no. They're not. They're much bigger, and I'm going to give you a sense of that.

Let's look at the picture here that says, "Weightlifting." First, we will look at the physiology of how I might—well, I'm not so much lifting the weight as I am simply holding the weight, so the picture here shows an arm, making a right angle bend, as I did when I was holding the bowling ball. We see the upper arm, and the lower arm.

If we want to get into the physiology, first of all we notice that the lower arm is holding on its end this massive object; in my case, the bowling ball. That's not part of the physiology, but that is there. Then, there's a structure of bones within the arm—the humerus, the upper arm bone, the radius and the ulna, the two lower arm bones—the skeletal structure is part of the physiology, but that's not enough to hold this bowling ball up. There is also a biceps muscle, and a tendon that attaches the biceps muscle to the bones of the forearm.

That's the physiology behind the very simple demonstration I just did, simply holding a bowling ball. Now, let's turn from the physiology to the physics.

As far as the physics are concerned, what we have is a system of basically, two rod-like objects held at right angles to each other, and at the elbow, there is a pivot joint, which is free to allow my lower arm to rotate about that joint. Rotational motion. That was the subject of Lecture Thirty-Two. We're going to start applying some concepts from Lecture Thirty-Two here.

First of all, the weight of that bowling ball is a downward force. It's the force of gravity on the bowling ball. There's also a downward force from the weight of the arm itself, but that's pretty insignificant in this case, compared to the weight of the bowling ball.

Now, that weight is a long distance from the pivot point, and remember from my demonstration in Lecture Thirty-Two with wrenches and so in that the torque—the twisting force, the force that tends to cause rotational motion to change—torque depends on the magnitude of the force, and how far that force is applied from the pivot point. It was more effective with a wrench that was longer than with a wrench that was shorter.

Well, that weight is a pretty big distance from my elbow, and consequently, it's a pretty large torque. There, I have shown you the distance from the elbow pivot to the weight. That's the distance involved in calculating the torque. You multiply that distance by the weight to get torque.

Now, my arm, when I was holding the bowling ball, was not in fact rotating. It was in a static situation. It was not undergoing rotational acceleration. Its rotational motion, which was to be at rest, was not changing, so there must not have been any net torque on it. Torque causes changes in rotational motion. There was no change in rotational motion, so there must be a torque in the other direction. This particular torque from the weight is tending to rotate my arm in this picture clockwise, or if I did the little right hand rule thing, into the screen. It must be a torque that is tending to rotate the arm counterclockwise, and that torque is provided by the only thing here in the picture that can provide the torque—that is, the biceps muscle, and the tendon that connects it to the forearm bones.

However, that connection point is only a tiny distance from the elbow pivot, and that means that the force in the muscle must be much, much, much larger than the weight. How much larger? Well, that depends on how big that weight is. That weight was about 15 pounds. The difference was about a factor of 10 between those two distances, and consequently, I may have had a force in my biceps muscle of 150 pounds. That's a pretty big force. Not as big as some forces that develop in one's body, though.

The forces involved in doing simple things like holding objects in our arms then, because of those principles of rotational motion and torque, tend to be larger than you might expect.

Here's another example. I'm going to jump off this stool. Now, our bodies, fortunately, know instinctively what to do to avoid breaking our bones when we do simple acts like jumping. If I were to jump perfectly stiff legged, and if The Teaching Company studio floor were not padded with carpeting, and made with pretty springy plywood, and I jumped stiff legged, for instance, onto a concrete floor, I would guarantee that I could shatter my bones. I instinctively, as I jump, avoid that problem. Let's take a look.

You see me bending my knees, and coming down. What's going on there? Let me do it once more. I'm jumping off the stool, and landing. I want to ask what kinds of forces were involved on my feet, and the bones in my legs, and so on, as a result of this simple act. Well, let's take a look at that. Here I have a picture called "Hard Landing." Here I am, standing on something. Here I am, jumping off, and landing on the ground.

How far did I jump? Well, I don't know, maybe 15 or 18 inches. Let's say that it was 18 inches, for example. I fell 18 inches. How long did it take me to stop? Well, that depends on things like how springy the floor is, how soft the carpet is, how springy the soles of my shoes are, and things like that. Let's suppose, as an example, that the stopping distance was an inch. It wasn't in the case of my jump. I'll show you that jump in a minute, pictorially, but suppose I jumped stiff legged, and had a stopping distance of an inch.

Now, what had to happen? I jumped off the stool, I gained energy from the gravitational force acting on me over that distance 18 of inches, the kinetic energy I gained turned out to be simply given by the force acting—my weight, times the distance, times the 18 inches. Then, the floor had to exert a force upward at me that took away that kinetic energy, so the product of that upward force, and the distance over which I stopped had to be equal to the product of the gravitational force and the distance I fell. Thus, to put it slightly mathematically, the weight times the fall distance has to equal the stopping force, times the stopping distance.

Well, the weight is my weight. The fall distance is 18 inches. The stopping distance is only an inch, and therefore, the stopping force is about 18 times my weight. For me, or for a typically sized person, that could amount to about 1.5 tons, so it's as if I were laying with my feet up in the air, and somebody dropped a 1.5-ton weight on my feet. There are very large forces involved in some of the commonplace things we do—or there could be if we didn't do them carefully, and instinctively acted to prevent danger to ourselves.

Once, in my early years as a physicist—I don't do this anymore—but I served as an expert witness in a medical malpractice suit that I won't go into. However, it involved forces on leg muscles and bones due to an accident that occurred in a hospital, and I spent an entire day on the witness stand with the lawyers from the other side trying to prevent me from stating the number, the number being the actual force that was exerted, because the number was so big that the jury would have been aghast.

Well, actually, although the number was big, and was enough to cause damage, there are, in general, very large forces exerted on our bodies in the regular, everyday acts of walking, jumping, and so on. They are just large, typically, as this 1.5-ton force, because what we do—as I did when I jumped off the stool—is protect ourselves. We undergo soft landings, so if I jump again, but just as I hit the ground, just as my feet which the ground, I bend my knees, the effect of that, to make the stopping distance for all of my body but my foot, it's a pretty low-mass thing, so it's not a big deal, there, the stopping distance is much smaller. Maybe it's nine inches, or something, depending on how much I bend my feet.

Well, it's still the case that the weight times the fall distance equals the stopping force times the stopping distance, and if the stopping distance is nine inches, and the fall was 18 inches, that means that the force on my feet was double my weight, about 300 pounds. That is certainly entirely tolerable.

We therefore do things instinctively, like bending our knees, to make a soft landing. By the way, if you want to look in detail at what is happening to a foot on a landing, here's a very high-speed photograph—a high-speed movie—just showing a foot landing on the floor, and you can see that rippling, wavelike motion going through the foot, as those forces are taken up by the soft tissue, transmitted into the bone, and so on. There are remarkable short-term stresses on our feet, as we do things like that.

By the way, there's a whole physics of automobile safety. Airbags in cars are doing exactly the same thing as my bending of my knees is doing what I jump off the stool. They're changing the stopping distance, making the stopping distance longer, so that the force is less dramatic.

That's an example of Newtonian mechanics in our bodies. What about electricity? Well, are nervous systems are good examples of an electrical system. We have cells called *neurons*, and they actually have extensions.

We think of cells as little microscopic things, but neurons have extensions that are up to several feet long, and electrochemical signals travel down those neurons. They travel relatively slowly, not anything like the near speed of light electrical signals in wires. That's why our response time is measured in tenths of seconds, for instance, rather than in microseconds, or something like that.

These neurons act very much like electrical cables, and some of the earliest understandings of neurons, when we really began to understand how the system worked, came from treating them as electrical cables, borrowing terminology from electrical engineers to try to understand how these worked. Simple circuit analogies apply to a lot of cases in the nervous system, and in other cases inside the body. I gave you an example way back in Module Three, where we look at the tiny currents flowing through the ion channels, bringing materials into the cell.

The body therefore has many, many electrical systems. One of the most obviously electrical, and most important, is the human heart. The heart is really very much an electrical system, and I'm going to give you a quick sense of how that works. I have a picture your showing a heart muscle cell. Heart muscle cells are elongated structures, and in their normal resting condition, they have a positive charge on the outside of that cell membrane, and a negative charge on the inside of the cell membrane. The charge is distributed all around. The cell is electrically neutral, and not only is it electrically neutral, but it doesn't have any of this asymmetric distribution of charge, like, for example, I've argued that the water molecule does.

During the beating of the heart, though, the cells temporarily depolarize. They lose some of this polarization across themselves, and here's how it looks. The polarized cell, because of the symmetry, has no significant electrical effect, but the heart muscle cell, when it's partly depolarized—not when it's fully depolarized, but partly depolarized—the cell has lost the charge on the inside and outside at the left. It now looks like a system in which positive charges slightly displace to the right of negative charges.

That's how I described the water molecule before. Water has oxygen negative, two hydrogens positive. It had a displacement of positive charge relative to negative charge, and even though it was electrically neutral, it still had electrical effects because of that. In the case of water, that's what made it a good solvent. That's what also allowed the microwave oven to grab a water molecule and vibrate it around.

Well, that's not what is happening to the heart muscle, but now the heart muscle does have electrical properties. They are the properties of the so-called *electric dipole*, a system in which charge is displaced slightly, but the whole system is neutral, and we represent the electric dipole with an arrow that points in the direction from the negativeness of the charge distribution to the positiveness. Thus, the individual heart cell becomes, basically, an electric dipole during the process of depolarization.

Now, let's look at the whole heart. This is a silly picture of a heart. It's sort of heart-shaped, and it's sort of tilted the way it would be in the body. It is full of these cells, and initially, the cells are all basically polarized. However, what happens as the heart beats is that a wave of depolarization sweeps down through the heart, so that the first layer of cells basically depolarizes. Now, things are more complicated than this simple picture, but this is a quick way of looking at the electrical effect of the heart.

Later, the second group of cells is depolarized, and so on. That wave of depolarization sweeps through the heart in a direction like that, and because all of those cells have the same electric dipole-ness, the heart acquires what a physicist would call, basically, an *electric dipole moment*, and what a cardiologist would call the *heart vector*. It's a vector that points, basically, in this direction from negative toward positive in the heart.

During the time this is happening, the electric fields associated with these dipoles, these separated charges, set up a field that is sort of the electrical analog of the field of a bar magnet, and if I plot what that field looks like—this is actual data on a human torso here—those contours are contours not of an electric field, but of equal values of an electrical potential of, basically, voltage, due to these dipoles in the human heart. When you get an electrocardiogram, electrodes are placed around the body, and basically, they're measuring these electric potentials.

When that happens, you get a signal that looks something like this. It begins with this slight rise. That's the *peak wave*. That's when the atria are polarizing. Then, there's a big peak in the middle, called the *QRS complex*. This is a normal looking heart electrocardiogram. That's when the ventricles are depolarizing. Then, in order to start the process all over again, the system has to repolarize.

What makes all this happen? Well, there are specialized cells called the *pacemakers* that initiate the electrical signals that get this going. When those go awry, and the heart starts fibrillating, for example, a fatal condition, you can stimulate it back into beating with the application of an electric current.

Well, that's electricity. What about fluid flow? I showed you a demonstration some lectures ago, in which I held what was basically a blower hose, and I supported a ball on that air stream, even if it was tilted. I argued that that was Bernoulli's principle, which says that when you have a low flow speed, you have a high pressure, and when you have high flow speed, you have a low flow pressure.

Here's an unobstructed artery, with uniform flow throughout, uniform speed and pressure. What if the artery builds up a little plaque, and gets obstructed? Well, then the blood flowing through has to speed up at the site of that plaque, in order to keep the same amount of blood flowing by. Higher speed, lower pressure. There is external pressure from your body, and the artery can actually collapse, a very dangerous situation, due to that lower pressure inside the artery, due to the higher speed caused by that collapse.

Those are some examples of physics in the operation of your body, and in this last case, in a possible problem your body could have. One of the major things we do with physics is to diagnose problems with your body. We've all had X-rays for our teeth, bones, and so on. X-rays were discovered by the German physicist Wilhelm Röentgen in 1895, in November. He quickly took an X-ray of his wife's bones, showed them to a medical society, and two months later, in January, X-rays were already being hailed as a medical miracle.

Modern X-rays can be as simple as what Röentgen did, which was to put a body part between a source of X-rays, and a film. I showed you how we made X-rays in a previous lecture, but much more sophisticated techniques include devices like CAT scans, in which X-rays are taken from a wide variety of angles. Here on the left you see a patient going into a CAT scanner, and the X-ray source moves around. The X-rays that go through are imaged, and the computer reconstructs a two- or three-dimensional picture of what's going on. On the right, you see a picture of a CAT scan of the brain, and these are good at basically revealing slices, two-dimensional slices, through the whole three dimensions of the brain.

A more sophisticated technique even than CAT scans, in some ways, is magnetic resonance imaging. That goes right into basic physics. It's based on a process called *nuclear magnetic resonance*. Here, I show a proton. A proton is a little magnet, because it is spinning. If I put the proton in a magnetic field, it will do what we discovered spinning things do back Lecture Thirty-Two. It will process. It can process with its spin approximately aligned with the magnetic field, or align the opposite way. It takes a little more energy to align it the opposite way, to push it against the field.

Here's a proton, spinning, with its spin aligned with the field. If a radio wave of just the right frequency comes along, that radio wave can flip the spin downward. Then, when the proton flips back up, it emits a radio wave of the same frequency, a little radio wave photon. That's basically how nuclear magnetic resonance works. How do we turn that into magnetic resonance imaging? By the way, the medical communities—physicians—are studiously avoiding the word *nuclear* here. There's nothing really nuclear going on, other than that you are working with atomic nuclei.

With magnetic resonance imaging, we deal with the protons, mostly in water and fat. They're in a magnetic field, and in an MRI machine, the magnetic field is non-uniform, which I have indicated here by the field being stronger on the right, with more of those arrows. Thus, only one of those protons is in the right place to respond to radio waves coming along. A radio wave comes along in the machine, and it misses that first proton, because it's not the right frequency to make it flip. The second one flips back, and gives us a signal, and that is the MRI's signal.

Because we know how the magnetic field varies with position, we know where that signal is coming from and the body. We can therefore reconstruct a picture, and the picture gives us a sense of what the density of the protons is in the body. That ultimately allows us to image structures.

Here's a picture of a patient being sent into that long, cylindrical magnet. It's usually a superconducting magnet that constitutes the core of the MRI machine, and on the right, you see a very detailed image of the brain, spinal column, features in the face, and so on. MRI imaging is very good at finding the subtle differences in the density of protons, and therefore, of different tissue structures within the body—a remarkable imaging technique.

Another way we image the interior of the body is by injecting or placing radioactive materials inside the body. An example, for instance, is radioactive phosphate, which gets absorbed by bones. Give the patient some radioactive phosphate, and then look at them with radiation detectors from outside. Here's a bone scan. You see the spinal column, again, showing up very nicely in this radioactive bone scan. A number of different isotopes are used. Iodine settles in the thyroid gland, so that you can use it for thyroid studies. Radioactive thallium is good for looking at blood flow through the brain, and so on.

Particularly interesting radioisotopes for looking at imaging inside of the body are PET scans, *positron emission tomography*. In my first job out of college, I was actually working on the process of PET, before it had become an active medical procedure. Here I show a nucleus of oxygen-15, which is radioactive version of oxygen, with only seven neutrons instead of eight; normal oxygen is 8 and 8, oxygen-16. This is oxygen-15. You have to make in the hospital in a cyclotron, because it's got only a two-minute half-life. It decays by emitting a positron, which is an antimatter particle, the antimatter particle of the electron. If you want to know more about antimatter, look at my course on modern physics.

Out comes this antimatter positron. It is in a universe where there are lots of electrons, not anti-electrons, and pretty soon, it encounters an electron, and the two get together, and they annihilate in a burst of energy, a 100% conversion of matter to energy, and out come two gamma-ray photons. They come in exactly opposite directions, and by $E=mc^2$, each has the energy of one electron's mass, or one positron's mass, because they are the same, and so we can detect these gamma-rays coming out, and we can look at where they have come from, with the rays of imaging detectors, and we can pin down exactly where this annihilation occurred.

You can have somebody breathe oxygen containing the radioactive oxygen-15, and you can actually watch as the oxygen goes into his or her lungs, and then is taken up by the blood. PET scanning is very good for processes where you want to observe things actually happening inside the body. These are procedures that are happening in real time. PET scanning is very good for that.

Here's an example of a PET scan. Here's a patient whose head is about to be put into the scanner, and at the right, you see a PET scan of the brain. Physics has therefore given us a number of quite dramatic techniques for imaging the interior of the body non-invasively commented in even mention one that I have talked about, way back in Lecture Six, which is ultrasound, and which, again, uses that partial reflection that occurs at a boundary of two different media, reflecting off internal boundaries between, say, organs, the bladder and the surrounding tissue, between the intestine and the surrounding tissue, and so on, to image with sound waves, going in and bouncing them off.

There are thus many, many ways to get inside our bodies essentially non-invasively. Sometimes, we put in dyes or radioactive materials, but a lot of the time, the techniques almost completely noninvasive. All we do with an

MRI is to get those protons spinning in a magnetic field, for example, and we get beautiful, detailed images of what is going on inside of you.

Well, physics also plays a role, as I indicated, in curing disease. How does that work? I talked about lasers. Lasers have come in many instances, replaced scalpels as basically precision knives for a variety of surgery. Way back in Module One, I talked a little bit about how laser eye surgery is used to reshape the cornea and correct your vision. However, laser surgery is used in a variety of other applications. In the eye, for example, it's used to weld a detached retina back to the back of the eye. A detached retina is a dangerous condition. When your retina detaches from the back of the eye, and begins to flow around in there, you can lose your vision that way. You can just tack that retina back down, and a little blast with the laser right through the eye can do that. Laser eye surgery is used basically to weld the retina back to the back of the eye.

Another advantage of laser surgery, as opposed to real lives, is that it leaves a lot less bleeding. As a laser knife goes through the proteins in the material of the body, tissues coagulate immediately, and reduce bleeding.

A process called *laser angioplasty*—angioplasty is a process of sort of reaming out your arteries. In laser angioplasty, a catheter is threaded into your artery, so that it's a little bit invasive. At the end of the catheter is a miniature high-powered ultraviolet laser, and when one gets to where there is a buildup of the fatty plaque on the artery wall, something that could break loose and cause a heart attack or stroke, that is blasted with the ultraviolet laser, and it's actually vaporized, so that laser angioplasty is a way of vaporizing the plaque that builds up in people's arteries.

If you have kidney stones, it turns out that one way to deal with kidney stones is to use shock waves, a concept I introduced in Lecture Six, when I talked about the shock waves produced as sonic booms by supersonic aircraft, for example. A shock wave, though, is really any extremely high intensity sound wave that has with it an abrupt pressure change. That abrupt pressure change exerts enormous forces on objects it hits, and so, using very high-energy sound waves, either from something that has been threaded into the body, or sometimes from a device that has simply been placed on the outside of the body, and focusing those sound waves down, one can blast kidney stones, and break them up to the point where they can be passed through the urinary system, for example.

If I have a broken bone, it turns out that electric fields, just associated with steady currents, can help make bones grow, so can ultrasound. So those techniques are used in cases where healing is difficult, or taking a long time.

Perhaps the most significant example of physics, though, in medicine and in curing disease is the use of radiation to cure disease, particularly cancer. There are medical physicists whose entire job it is just to figure out what dose of radiation one needs to do that. Here's basically why radiation works. Radiation is particularly effective against cells that divide rapidly, which characterizes cancer cells. That's also, by the way, why children, whose cells are dividing more rapidly than are adults, are particularly susceptible to radiation dangers.

Cancer cells are also less organized, and most cells have repair mechanisms that actually let them repair the damage caused by minor radiation doses. Well, the less organized cancer cells aren't as good at that, so they are more susceptible to radiation.

Here's a graph showing the percent of cells killed on the vertical axis as a function of the radiation dose given. For healthy cells, it might look like that. Give a little bit of radiation and not much happens because the cells can repair themselves. After a while, you reach a point where, if the radiation dose increases a little bit, the mortality of those cells goes up significantly. Finally, at a big enough radiation dose, most of the cells are killed.

That's for normal cells. Cancer cells, because they're dividing rapidly, and because they're less organized, and don't have effective repair mechanisms. The high mortality kicks in earlier, so the name of the game in radiation therapy is to look at these two curves, figure out what kind of cancer cells you've got, and so on, and find a dose that maximizes the survival rate of the patient. For example, if the patient has a cancer that is definitely going to be fatal, if you give them no radiation, they're going to die. If you give anybody, healthy person or cancer patient, enough radiation—it takes a big amount, by the way, for it to be fatal, but it has happened; we know the amounts, roughly, from studies of the Japanese nuclear bombing survivors, and the fairly rare cases of nuclear accidents—if you give big enough radiation dose, the patient is going to die. Period.

Therefore, it is somewhere in between—and it is somewhere at the point where healthy cells are beginning to be killed, but not in great big numbers, and cancer cells are being killed in pretty big numbers—that's the optimum point. If you can achieve that dose of radiation, you can enhance the survival rate of the patient to the maximum. Not to 100%, unfortunately, but too maximum.

Early radiation therapy used gamma-rays, actually emitted by radioactive material, radioactive cobalt, in particular, but that was difficult to control: Difficult to control the penetration, depth, and so on. Most modern radiation therapy uses miniature accelerators, sort of like miniature atom smashers that accelerate electrons or protons, high-energy beams of particles, and the intensity of those beams can be controlled. Protons, although they have only coming to use largely in recent years, are good because they deliver their energy more at the site of the tumor, that you are trying to get rid of, rather than throughout the body. Of course, you want to minimize damage to healthy cells. That's one effect of this physics of collision that I've talked about, back in Module Two.

In radiation therapy then, we put a patient under a machine, basically, in the beam of a high-energy particle accelerator, and bombard him with these particles. Sometimes we do so with beams coming from a lot of different directions, which maximizes exposure of the tumor, and minimizes the exposure of other tissues.

Physics is a very big part of our lives and of our bodies, and particularly, it can cure us of some diseases.

Lecture Thirty-Six
Your Place in the Universe

Scope: You are the product of an evolutionary sequence that began with the simplest possible states of matter and energy and proceeded through ever increasing complexity to the point where at least one agglomeration of matter—the human brain—became conscious and able to contemplate the Universe around it. A remarkable set of subtle circumstances allowed the Universe to evolve the chemical elements that proved necessary for life as we know it. In the early stages of cosmic evolution, only the simplest elements— hydrogen and helium—were produced in significant quantities. All the other elements, common and uncommon, up to and including those as heavy as iron, were formed in the interiors of long-dead stars. Those stars exploded, spewing their contents into interstellar space. This "stardust" eventually condensed to form new stars and planets. On at least one of those planets, life arose and, eventually, consciousness. Our everyday lives are, thus, intimately connected with the lives of the stars and with the earliest instants of time.

Outline

I. Your place in the Universe is planet Earth, a rare chunk of solid matter that is much cooler than the visible matter of the Universe but much warmer than the overall average temperature. Yours is the place where the Universe began—but so is every other place. Your place in time is some 14 billion years since the Universe began. You're connected to cosmic history through the matter of which you're made.

II. A brief history of the Universe shows an evolutionary sequence beginning with the simplest forms of matter and energy and evolving through ever more complex structures, including some 14 billion years after the beginning, the conscious brain. The dominant theme in this evolution is the expansion of the Universe and the concomitant cooling—a process that lowers the average energy and allows more complex structures to coalesce and persist. Our detailed understanding of this cosmic history depends on many disparate observations, but the overall picture comes from three distinct discoveries.

A. In the early 20th century, the Universe was believed to be essentially static. Individual stars and planets might come and go, but the overall structure had always been the same and always would be. So ingrained was this belief that it led Albert Einstein to commit what he called "my greatest blunder." Einstein's 1916 general theory of relativity suggested that the Universe should not be static but expanding or contracting. Einstein introduced a "fudge factor" called the *cosmological constant* to provide a kind of "antigravity" that would maintain a static Universe against the mutual gravitational attraction of all its components.

B. In the 1920s, the astronomer Edwin Hubble (for whom the space telescope is named) made a series of observations that showed the light from distant galaxies to be reddened, with the degree of reddening increasing with distance.

    1. Hubble used the then-new 100-inch telescope at Mount Wilson outside Pasadena, California. Before the advent of this telescope, there was controversy about the very existence of other galaxies. Today, we know the galaxies as vast agglomerations of stars and other matter, containing typically tens to hundreds of billions of stars.

    2. Hubble interpreted the reddening of the distant galaxies' light as a Doppler shift (recall Lecture Six). Because reddening represents a lengthening of light's wavelength, Hubble concluded that galaxies were moving away from us with speeds proportional to their distances. Hubble's discovery suggested that the Universe is expanding, with galaxies rushing away from each other in the aftermath of a cosmic explosion—the *Big Bang*—that began a finite time ago. Although it might seem like Earth is at the center of this expansion, it isn't. There is no center or, equivalently, every point has equal claim to be the "center."

C. The idea of an evolving Universe remained controversial through the mid-20th century. An alternative, the *steady-state* theory, suggested that matter was continually created to fill the void left by the receding galaxies. That way, the large-scale structure of the Universe would remain unchanged. But then a 1965 discovery gave overwhelming evidence for a Universe beginning with a Big Bang. In that year, two Bell Laboratories scientists, Arno Penzias and Robert Wilson, were trying to find the source of some "noise" (static) in a radio antenna. They found that the noise came from outside their antenna and seemed independent of where the antenna was pointed. Coincidentally, theorists at nearby Princeton University predicted that a Universe starting in a Big Bang should, today, be pervaded by microwave radiation with just the characteristics being measured at Bell Labs. Penzias and Wilson had discovered the *cosmic microwave background radiation* (*CMB*). Today, the CMB provides some of the most precise and detailed information we have about the early Universe. More on this topic shortly. (Penzias and Wilson shared the Nobel Prize for their discovery.)

D. Through the 1970s and 1980s, both theorists and experimentalists working with large particle accelerators made advances in our understanding of elementary particle physics. (Much more on this in Professor Pollock's Teaching Company course *Particle Physics for Non-Physicists*). These new understandings were then applied to explore the interactions of matter and energy in the very early Universe and to predict the formation of simple structures, such as nuclei and atoms. Today, particle physics, which deals with the smallest objects in the Universe, and cosmology, which deals with the largest scales, are inextricably linked.

III. The timeline for cosmic evolution starts a mere 10^{-43} of a second—that's 1 over 1 followed by 43 zeroes—after the beginning. We can't go back further than that because our present understanding of physical reality does not cover conditions that prevailed before that time. To do so would require a merging of quantum physics with general relativity, and that has yet to be achieved.

A. At the earliest times we can know, the Universe was an almost structureless soup of matter and energy. Only the most basic of particles could exist, and the temperature was far too high for them to stick together. Creation of matter from energy and annihilation of matter into energy occurred routinely. Although this description has been on good theoretical footing for some time, experiments in the early 2000s at Brookhaven National Laboratory have momentarily reproduced matter under conditions that existed before even a millionth of a second had passed.

B. Around a millionth of a second out, the temperature had dropped to the point that quarks could join to form the familiar protons and neutrons that we know from nuclear physics. But these particles still couldn't stick together.

C. By about 3 minutes, the temperature had finally dropped to the point where protons and neutrons could join to make atomic nuclei. For roughly the next half hour, protons and neutrons joined to make helium nuclei (2 protons, 2 neutrons) and trace amounts of a few other simple nuclei, notably deuterium (hydrogen-2; 1 proton, 1 neutron; recall Lecture Thirty-Four). At the end of the first half-hour, the temperature and density were too low for further *nucleosynthesis*, and the process stopped. Theoretical calculations show that the matter of the Universe at that time should have been about 25 percent helium, 75 percent hydrogen (bare protons). When we observe the Universe at large today, we indeed find that it consists of about 25 percent helium and 75 percent hydrogen. The rest—including the elements we're made of—is a tiny deviation from this hydrogen-helium Universe. And virtually none of the elements beyond hydrogen and helium were formed in that first half-hour.

D. At about 300,000 years, the temperature had dropped to the point where electrons could join nuclei to form neutral atoms. Before this, the Universe had been opaque, because free charged particles readily absorb light (a form of electromagnetic radiation; recall Lecture Eighteen). But at 300,000 years, the Universe suddenly became transparent because it now consisted largely of neutral atoms. The electromagnetic radiation emitted as electrons "fell" into orbit around the nuclei was therefore free to travel throughout the Universe with little chance that it would be stopped by matter. That radiation would become the CMB.

    1. As the Universe continued to expand and cool, the wavelength of the radiation itself lengthened, until today, it is predominantly microwave radiation. The distribution of the radiation is characteristic of the average temperature of the Universe, about 2.7 kelvins, or 2.7 degrees Celsius above absolute zero (recall Lecture Twenty-Five).

    2. Satellites designed explicitly to study the CMB have confirmed that it is remarkably uniform, with essentially the same intensity and temperature in all directions. But there are tiny fluctuations, and they tell us a great deal about the structure of the early Universe. In particular, the fluctuations indicate a "lumpiness" that represents the "seeds" of the large-scale structure—galaxies and clusters of galaxies—that characterize the Universe today. And detailed analysis of the sizes of the CMB fluctuations gives evidence for a period of exceptionally rapid expansion that occurred around 10^{-32} second.

3. You, too, can observe this "fossil" radiation from the beginning of the Universe. Tune your TV to a channel where there's no signal in your area, and you'll see "snow" on the screen. About 1 percent of the energy being captured by your TV to make this snow is the CMB!

E. When the Universe was a few hundred million years old, the slight "lumpiness" evident at the time the CMB formed led to gravitational "clumping" of matter to form galaxies. With new instruments mounted on the Hubble Space Telescope, we can see distant galaxies as they were a mere 400 million years after the beginning and, thus, begin to understand how galaxies form and evolve and how they're distributed in the Universe.

F. For the first roughly 9 billion years, the Universe continued to expand at an ever slower rate, as the mutual gravitational attraction pulled at the receding galaxies. But about 5 billion years ago, the expansion began to accelerate. This phenomenon, discovered only in 1998, shows the existence of a mysterious force, called *dark energy*, that effectively counters gravity. Understanding dark energy is an ongoing goal of contemporary cosmology.

IV. New observations of the early 2000s show that the first stars also formed just a few hundred million years after the beginning, perhaps in a burst of extravagant star birth. Since then, the stars have continually recycled and processed the material of the Universe, building up a mixture richer in elements heavier than the primordial hydrogen and helium.

A. Stars form as gas and dust agglomerate under the influence of their mutual gravity. Regions dense in gas and dust constitute "stellar nurseries" with prolific star formation.

B. As more matter accumulates in the nascent star, the pressure and temperature at the center increase. Eventually, the temperature becomes high enough for hydrogen nuclei to begin fusing together (recall Lecture Thirty-Four). The immense gravity of the stellar material confines the fusing gas, and the star settles into a long period of steady "nuclear burning."

1. How long a star burns its nuclear fuel depends on its mass. The Sun has a lifetime of about 10 billion years; 5 have already passed. But a star with 30 times the Sun's mass only lives about 10 million years.

2. A star begins with a series of nuclear reactions that ultimately convert hydrogen (a single proton) into helium-4 (2 protons, 2 neutrons). In more massive stars, three helium nuclei can fuse to make carbon. Actually, two first fuse to make an unstable isotope of beryllium. By a wonderful fluke of nature, the beryllium lasts just long enough that a third helium nucleus can join to make carbon. If that did not happen, there would be no heavier elements and no life!

3. As a massive star exhausts the hydrogen at its core, it collapses and heats up. Then helium "burning" begins. Once helium is exhausted, carbon "burning" commences, producing a myriad of different nuclei—most commonly oxygen and silicon, which just happen to be the most abundant elements on Earth. The process of collapse, heating, and new nuclear reactions continues through the production of iron, the most stable nucleus (recall Lecture Thirty-Four and the curve of binding energy). Near the end of its life, a massive star has an onion-like structure with layers of different elements formed in the various fusion processes. The abundance of the various elements in the Universe is established by these nuclear processes.

4. Unlike our Sun, which will end its life rather calmly, a massive star ends in a violent *supernova* explosion that spews the star's contents into interstellar space. Eventually, the material is recycled into new stars, and the process repeats— each time enriching the Universe with heavier elements. (Elements heavier than iron don't form in the stars but in the brief moments of the supernova explosions themselves.)

C. When new stars condense from the interstellar medium, they may form stellar systems, with less massive planets in orbit around a central star. The first extrasolar planet was discovered in 1995, and today we know dozens of planetary systems. Our solar system, which formed some 5 billion years ago, is not unique.

1. Less massive planets have weaker gravity and cannot hold onto such light elements as hydrogen and helium. Thus, they develop into solid bodies rich in such elements as oxygen, silicon, carbon, and iron.
2. At a range of distances from the central star is the *habitable zone*, where a planet with the requisite size and composition would have an environment conducive to the origin and evolution of life. In our solar system, that zone encompasses Venus, Earth, and Mars, but at present, only Earth has conditions conducive to life.
3. Sometime around the first billion years of Earth's existence, life arose. Soon, photosynthesis by blue-green algae began changing Earth's atmosphere, introducing a significant level of oxygen.
4. Roughly a billion years ago, multicellular organisms evolved.
5. Some 400 million years ago, life began to colonize the land.
6. Sixty-five million years ago, an asteroid collided with Earth, wiping out the dinosaurs and paving the way for mammals.
7. Somewhere around 4 million years ago, the first humans evolved, with a consciousness that let them understand and appreciate the Universe and, eventually, its long history. That is our place in the Universe today: literally, descendants of the stars and the nuclear processes that forged the elements that today make our bodies, brains, and ultimately, our consciousness.

V. What of the future? In a paper entitled "Time without End: Physics and Biology in an Open Universe," physicist Freeman Dyson argues that, once established, consciousness can exist into the infinite future of an ever-expanding Universe. Recent studies, especially the 1998 discovery of accelerated expansion, somewhat temper Dyson's optimistic assessment. But it remains a grand vision of physics in our lives and our lives as part of the ongoing physics of the whole Universe!

VI. This course has been about physics—mostly practical, everyday physics. Yet physics touches all other aspects of our lives; for this reason, I end, fittingly, with literature: selected passages from Pattiann Rogers's poem "The Origin of Order."

Suggested Reading:

Lawrence Krauss, *Atom: An Odyssey from the Big Bang to Life on Earth...and Beyond.*

Sky and Telescope, October 2003, special issue, "The New Cosmology."

Going Deeper:

Eric Chaisson, *Cosmic Evolution: The Rise of Complexity in Nature.*

Alan Guth, *The Inflationary Universe.*

James B. Kaler, *Cosmic Clouds : Birth, Death, and Recycling in the Galaxy.*

Questions to Consider:

1. In what sense is the CMB radiation a "fossil" from an earlier time in the history of the Universe?

2. Trace your own lineage back to the core of a long-dead, massive star.

Lecture Thirty-Six—Transcript
Your Place in the Universe

Welcome to Lecture Thirty-Six, the final lecture, "Your Place in the Universe." In the previous lectures, I've tried to give you a sense, sometimes more theoretically, sometimes more practically, of the many, many roles physics plays in your life, but there's a very deep, fundamental role that physics plays in your life, and that is because you yourself, the materials which your body is made, the evolutionary sequence that has led to the intelligent being that you are—those are all products of a remarkable cosmic evolution that's been going on for something like 14 billion years, and you would not be here if it were not for a lot of detailed physics that went into that evolution.

I want to take a brief time in this final lecture to map out that cosmic evolution, ending up with you as an intelligent being. This is really the stuff of some other courses, particularly, Professor Filippenko's course, Professor Tyson's course, and to some extent, Professor Pollock's course, the first two on cosmology and astronomy, and the last one on particle physics. They have a lot more to say about these matters than I'm going to say here.

What is your place in the Universe? Well, in terms of "place," literally interpreted, your place is on a relatively cold chunk of rock we call planet Earth. It's much cooler than most of the visible material we see in the Universe, the stars. On the other hand, it's much warmer than the average temperature of the Universe, which is only 2.7 Kelvin.

What about your place in time, which also means part of your place in the Universe? Well, your place in time is roughly 14 billion years or so after what we think was the beginning of the Universe, and the Universe has had a beginning, an evolutionary sequence that began with the simplest forms, and whose main theme was the idea that as the Universe gets bigger and cools down, it becomes more complex, more and more complex structures can form, and I won't say it's the final endpoint, but where we are today, here on Earth, what we know of as the human brain is probably the most complex and remarkable structure that has evolved in that evolutionary sequence.

The dominant theme: expansion and cooling, ever more complex structures, coalescing out of what had been the primordial soup. Now, we haven't always had that vision. In the early 20th century, people assumed the Universe was completely static. It always looked pretty much as it has, as it does, as it always has, and always will. It had been there forever. In fact, so prominent was that view that Einstein, when his general theory of relativity came out in 1916, was embarrassed to find that his general theory predicted that the Universe should be either expanding or contracting, and he knew it wasn't, because that was the prevailing view. He therefore introduced a "fudge factor" into his equations so that it wouldn't be viewed as expanding. He later called that his biggest blunder, although in 1998, we became a little less clear as to whether or not he was wrong. I'll get to that later.

In the 1920s, though, after the construction of the 100-inch Mount Wilson telescope, the biggest telescope of its time, faint, fuzzy patches that had been a matter of debate—were they in our own stellar system, or were they some distant island Universes—it became clear that they were, in fact, distant—"Universes" isn't the right word. They were distant islands of stars, maybe billions of stars, like our own Milky Way—the galaxies had been discovered.

In a series of observations in the 1920s, Edwin Hubble observed distant galaxies with the Mount Wilson telescope, and he discovered a remarkable relationship. He noticed that galaxies that were far away from us seemed to be reddened. Their light was reddened, and the further away they were, the redder the light was. Now, redder light means longer wavelength, as we know, and the best interpretation of that—the simplest interpretation of that, and the interpretation, we believe is correct—is that the reddening is due to the Doppler effect. Remember that longer waves get Doppler shifted to longer wavelengths if the source of the waves is moving away from you. Thus, with the exception of the very nearest neighborhood galaxies moving sort of at random with respect our galaxy, all the galaxies in the Universe are moving away from us at speeds that depend on their distance.

That's called the *Hubble relation*, and here's a simple graph of it. On the left, on the horizontal axis, are nearby galaxies, on the right are faraway galaxies, on the vertical axis. This is the speed at which they are receding from us. If you measure some different galaxies, you get a little bit of scatter as to how these points lie, but they basically describe a straight line, and that straight line is the Hubble relationship. It's the relationship that says, "Okay, the further away you are, the faster you are going."

What is the interpretation of that? Well, one interpretation is to think of the current picture of the Universe as a kind of movie that we might run backwards, a frame in a movie that we might run backwards. Why are the farthest galaxies farthest away from us? Because they are moving the fastest. At some point, the galaxies were all in one place, and we can extrapolate backwards in time, and ask when that point was. The answer comes out—depending on exactly what the value of the Hubble relationship in the Hubble constant is, the slope of this line—somewhere between, typically, 10 and 20 billion years ago. We have now pinned down that number, with considerably more accuracy, to 14-something billion years ago.

That's the origin of the idea that the Universe began with a "Big Bang" event. It sounds awfully like we are at the center of the Universe, but we are not. I'm going to give you a simple example. Here's example involving raisin bread. Imagine a big blob of raisin bread. This is a picture of a blob of raisin bread, and each of the black dots is a raisin, and it's expanding.

By the way, in my lecture on lasers, Lecture Thirty-Three, I sent some signals over a laser beam, and those signals were in fact some verbiage from my course on modern physics, and I was talking about exactly this example, about the raisin bread analogy for the expanding Universe, possibly thinking about an infinite loaf of raisin bread, but it's a finite loaf. Here's a picture of it.

Here are a couple of galaxies that are some distance apart, and later, after the Universe has expanded, they're now a bigger distance apart. Here are a couple of other galaxies. They're some distance apart. Later, after Universe has expanded, they are a bigger distance apart, and so, the first pair of galaxies were some distance apart, then, they were a bigger distance apart, and the distance grew by a certain amount.

At the same time, those galaxies that started out more distant were some distance apart, then a bigger distance apart, and the distance grew by a certain amount. At the same time, those galaxies that started out more distant grew by a bigger amount, and that's an explanation of the fact that as the Universe expands, the galaxies that are further apart end up moving faster. They have to be a cover up the same overall expansion in a given time. You might still say, "Well, that means we're at the center," but we're not because if you are in this loaf of raisin bread, every raisin sees the same thing. If you are at that one raisin that doesn't have an arrow on it, you see all the other raisins moving away from you, and they are moving at speeds that depend on their distance.

If you go to another raisin, it sees exactly the same thing, and the only problem with this analogy is that the raisin bread is finite, so you have to think about an infinite loaf of raisin bread, and that's a little bit mind-boggling. Another way to think of it is to imagine something like the surface of this balloon. I got a simple line here, and I've marked on it red dots that represent galaxies. I'm going to blow it up, and what's going to happen is that every galaxy is going to move apart from every other galaxy as the Universe, in this case, the two-dimensional surface of this balloon, expands, and yet, no galaxy is at the center. The down inside the middle of the balloon isn't part of the Universe. This is a two-dimensional analogy that's going on in our three- or four-dimensional Universe, the dimensions of space and time. It's expanding, and every galaxy is moving apart from every other one, but no one is at the center. Here goes a balloon analogy for the expanding Universe.

As the Universe expands, each galaxy moves away from other galaxies, and the speed of that motion depends on how far apart those galaxies are. Those two galaxies didn't have to move very fast to get far apart in that time. Two galaxies further apart, those two had to move faster. The galaxies move apart with recession speeds that depend on their distances from each other. That's the expanding Universe.

Now, until the mid-20th-century, this idea of an expanding Universe was controversial—of an expanding Universe with a beginning. The idea of the "Big Bang" was controversial. There was a competing theory called the *steady-state* theory. The steady-state theory said, "Yes, the Universe is expanding, but as it expands, matter is created out of nothing to keep the overall density of the Universe the same." It sounds odd, but it was needed to avoid the problems associated with saying that the Universe had a beginning.

Then, in 1965, there was a very important discovery made. It was made by a couple of scientists at Bell laboratories, Arno Penzias and Robert Wilson. They were cleaning out a radio telescope, which had some static in it, some noise, and they were trying to figure out what the source of that noise was. There were some pigeons nesting in the telescope, and they thought that perhaps it was pigeon dung, so they shoveled that out, they cleaned it out, and they couldn't get rid of the noise. It was coming from all directions, and this was at Bell laboratories in New Jersey.

Coincidentally, theoreticians at Princeton University were studying what the effect of a Universe with a "Big Bang" would be, and they argue that there ought to be, now, a sea of microwave radiation pervading the entire Universe, characteristic of an object at a temperature a little below three Kelvin. Those two groups of scientists heard about each other, a link was made, and it was recognized that the Bell Labs scientists had studied this *cosmic microwave background radiation*, as it's called, which was a sort of fossil remnant of a time much earlier in the Universe.

I talked about this briefly in a couple of earlier lectures; and today, this cosmic microwave background provides some of the most detailed knowledge we have about the history of the early Universe. I'll get back to that a little bit later.

Then, in the early 1970s and '80s, just after the discovery, particle physicists, working with the structure of high-energy particles, and their particle accelerator laboratories, were beginning to understand the reactions of elementary particles under the kinds of conditions that would have existed early in a Universe that began with the Big Bang. You can find a much more about that, again, in Professor Pollack's course, *Particle Physics for Non-Physicists*.

These new understandings from the microscopic realm of particle theory physics, and from the realm of cosmic microwave background radiation, were put together to give a picture of what the early Universe might have looked like, based on what we know theoretically. Today, we start the cosmic timeline at about 10 to the -43 seconds. That's a one over one with 43 zeros after it. We can't go back further than that, and the reason is that at that point, the Universe was so hot and dense, and so compactly crammed into one place, while—maybe it was still infinitely big; if it's infinite, we can argue about that. It's still not clear, but it begins to look like it's infinite.

Conditions were such that you needed both general relativity and quantum physics to explain what was going on. We simply have not—as I said in the first lecture—yet figured out how to join those two theories. We therefore have to start at this point at about 10 to the -43 seconds. We just can't go back any further, because the physics we have don't allow us to; and we know at that at that point, the Universe was basically an undifferentiated soup of matter and energy. Processes were occurring—namely, the creation of particles out of pure energy, and the annihilation of particles back into pure energy, the process that was involved in positron emission tomography, that I discussed in the previous lecture.

That's a rare process now, but that was going on in the Universe. The number of particles was not fixed. Things were running back and forth all the time, because there was so much thermal energy in this hot, dense stuff that was the early Universe.

The basic particles, if they banged into each other, the energy was too high for them to stick together and form anything more complex. You might say, "Well, how do we possibly know about this?" We were on pretty good theoretical footing in this explanation of the early Universe for some time, but at around the year 2000, experiments at Brookhaven National Laboratory began to actually recreate conditions that existed in the first millionth of a second or less of the Universe, and bear out that our understanding of the particle physics of that time was probably correct.

The picture I'm showing now is a picture of a spew of particles coming out in the collision of two nuclei of gold, that in this Brookhaven machine, the relativistic heavy ion collider, so called because these gold nuclei—and gold is a pretty heavy nucleus—are accelerated to almost the speed of light. Two gold nuclei bang into each other head on, and out comes a spew of elementary particles. From that, we can understand, momentarily, what conditions were like in the early Universe.

This is well before a millionth of a second. At about a millionth of a second, the temperature had dropped to the point where quarks—there were quarks and electrons, basically; those were the really fundamental particles that existed—where quarks could begin to stick together to form the familiar protons and neutrons. The protons and neutrons couldn't stick together yet, but quarks could stick together, to join individual protons and neutrons.

However, as the Universe expanded and cooled—and here's this big theme: expand and cool—more complex structures formed. Now, we had nucleons, protons and neutrons. Before one millionth of a second, we didn't have those. At about three minutes, the temperature dropped to the point where those protons and neutrons could finally begin to stick together, and for the next 30 minutes, what an important half-hour in the history of the Universe. You've been through 36 half-hour lectures, but that half-hour capped them all. That half-hour was a half-hour in which protons and neutrons joined, and some of them joined protons and neutrons to make deuterium, that second isotope of heavy hydrogen.

Some of those deuteriums joined to make two protons and two neutrons, helium, and there was a smattering, just a smattering, of a few other things: A little bit of lithium formed, for example, deuterium, as I mentioned, but mostly what happened is that for half an hour, these nucleons joined, and they made helium.

At the end of that half-hour, things got so undense that that process was unlikely to occur anymore. Consequently, at that point, we had a Universe that the theoricians predict should have been about 75% protons— hydrogen—and about 25% helium, and a tiny, tiny trace amount of just one or two other elements, such as deuterium, a version of hydrogen, and lithium, for example. Almost nothing else.

We look out in the overall Universe—not at planet Earth, not at our Sun, even, wherever we think we can see primordial stuff, we sort of average over everything we see in the Universe—and in fact, the Universe we observe consists of about 75% hydrogen, and 25% helium, a nice confirmation of the predictions of this Big Bang theory.

Now, at some time that I've described earlier this course as roughly half a million years out, 500,000 years out, and probably more accurately, about 300,000 years out, we still had just these atomic nuclei, protons and helium nuclei, and electrons were whizzing all around, but the electrons—again, there was still too much thermal energy for the electrons to join, to make atoms. However, about 300,000 years out, it got cool enough that they joined with protons, and could make hydrogen atoms, or helium atoms, when they joined with helium nuclei.

Before that, the Universe was opaque to light and other electromagnetic waves, like radiation, because, again, those particles were individually charged particles, and they interacted strongly with the electric field of electromagnetic waves. After they formed atoms though—which are neutral—there was much less interaction. There was still some, but there was much less interaction with electromagnetic waves, and at that moment, the Universe became transparent.

The radiation that was emitted as the electrons fell into those orbits and made the atoms—we suddenly saw a transparent Universe—and that radiation could travel throughout the Universe, basically forever, with relatively little chance that it would ever bump into anything—that's the cosmic microwave background radiation.

At the time it was formed, it was visible ultraviolet and higher energy radiation, but as the Universe expanded, the expansion of the Universe dragged out the wavelength of those waves, until today, their characteristic of the distribution of wavelengths in an object is only about 2.7 Kelvin above absolute zero, so that's the number that we described as the temperature of the Universe. We study that cosmic microwave background today with satellites, and a first glimpse shows that it looks remarkably uniform. This is a picture of the sky, a sort of map of the sky, as observed from a satellite, and this is the intensity of the cosmic microwave background. It looks almost uniform.

It has variations, though, to one part in about 100,000, which gives us details about the lumpiness of the Universe at that time, 300,000 years out; and in fact, some of this information takes us back to before a second, if we interpret this correctly, because we actually see residues of sound waves that were bouncing around in the early Universe. This is a picture from the Wilkinson Microwave Anisotropy Probe, a spacecraft, by the way, that is parked at the L-2 Lagrangian point. I described for you the L-1, .5 million kilometers, one million miles one side of Earth. There's another point a million miles the other side of Earth, and it's a nice place to put a spacecraft looking out into deep space, because the Earth, Moon, and Sun don't get in the way.

By the way, if you would like to observe the cosmic microwave background, you can do so. Just turn on your TV, and tune it to a place between channels. It's much more interesting than what you see on most of TV, and look at what you see. You see that snow bouncing around. That snow is random noise due to light switches going on and off, random microwave interference, microwave ovens, and all kinds of stuff like that. However, about 1% of it is, in fact, the cosmic microwave background radiation, so you yourself can observe the cosmic microwave background radiation on your TV.

Now, I showed you these pictures that suggested that there was a little bit of lumpiness in the Universe even 300,000 years out. That lumpiness as the Universe expanded led to gravitational agglomerations of material, which ultimately led to the formation of galaxies. Recent observations by Hubble suggest that galaxies actually began forming within a few hundred million years of the Big Bang. By 400 million years, there were probably galaxies, and today, we have detailed studies going on of the distribution galaxies at the farthest reaches of space that we can see.

Here, for example is a picture of what is called the *Hubble Ultra Deep Field*, a very longtime exposure at the Hubble station's telescope, designed to bring out galaxies at the farthest reaches of our vision, which is to the edge of the observable Universe, we think, almost. You see galaxies of all sorts, of all distances, of all colors, of all shapes, strewn about here. Each of these galaxies has about 100 billion stars.

When you look in the depths of space, don't look in a densely star populated region of the Milky Way, but out into the depths of intergalactic space; there are galaxies everywhere. There are great, huge numbers of galaxies, enormous numbers, and in fact, here's a survey. This is from the Sloan Digital Sky Survey, which shows on the right, a small picture of a slice of space. Some of the nearby objects are stars, but most of the objects are galaxies, and by studying those slices of the sky, a three-dimensional map is constructed of the distribution of galaxies. This particular study looked at some 200,000 galaxies, and in the picture on the left, you see the distribution of those galaxies in space. That's the distribution of only the 6,000 or 7,000 nearest galaxies.

This expansion continued for about the first nine billion years, the Universe continued to expand, but the expansion slowed, and the reason it slowed, we believe, is because all that matter, which is rushing apart from each other, is, after all, exerting gravitational force on itself and on each other, and slowing the expansion. It's just like when I throw a ball, it slows down. The question has long been, "Will the Universe ever slow down and come back?" We think we've answered that question now, and the answer came back very surprising.

In 1998, studies of distant supernovas, now corroborated by astronomical observations, suggested that about five billion years ago, coincidentally, at about the time the solar system, the Sun, and the Earth were forming—but that's pure coincidence—the slowdown stopped, and the expansion began to accelerate. That's evidence for some kind of new, unknown energy, force. It's called *dark energy*. It's kind of anti-gravity, that is acting to push these things apart.

We don't understand what that is, and as I was going over my notes for this lecture this morning, an email beeped on my screen from the American Astronomical Society, and it contains news of interest to the astronomical community, as they send out every so often. One of the news items was a decision by the National Aeronautics and Space Administration, NASA, and the US Department of Energy to make the investigation of dark matter of very high scientific priority in the coming decade. This is something we need to understand. This is part of that 95% of the Universe that we do not know what it's about.

Well, that's the big picture, but now, let me zoom in, because I want to get, in the last few minutes of this course, to the physics of the Universe in your life. In the early 2000s, new observations showed that stars were beginning to form as early as 100 million years, a few hundred million years, after the Big Bang event. What stars do is basically build new material and recycle material. Stars form out of agglomerations of gas and dust. I have a picture here from the Hubble space telescope—a gorgeous picture of a big cloud of interstellar dust, and you can see kind of glows coming around it. That's from new stars embedded in that cloud that have just been forming, new, young stars that are glowing very brightly.

Stars are continually forming in the Universe in our own galaxy, and astronomers are actively studying those star formation regions. As more and more matter accumulates in a new star, the density grows, the pressure due to gravity grows, and eventually, the star becomes hot enough in the center that there are nuclear fusion reactions, as I discussed in Lecture Thirty-Four, starting to take place. At that point, the star is sustained against further collapse by the nuclear fusion reactions, and the star settles into a sustained period of steady so-called *nuclear burning*.

How long does it burn? Well, that depends on the mass of the star. A star like our Sun, which is a fairly puny everyday star, will live about 10 billion years, and it's about halfway through its lifetime. It's a middle-aged star. However, a star that's much more massive, like a star that's 30 times the mass of the sun, will only live 10 million years. The rate at which those nuclear reactions go, grows very rapidly with the mass of the star, and it's those stars that are particularly interesting in this story.

What happens in the star? In a star like the Sun, hydrogen fuses into helium. A star makes helium. Good. In the Sun to some extent, but more importantly in more massive stars, there are reactions that then take helium and fuse it into other things. Here's a remarkable coincidence. If three helium get together—helium is two neutrons, and two protons—if three get together, they make six neutrons, and six protons, and that's carbon, the element essential for life.

Well, it's very unusual that three particles will collide all at once, a very rare event. What if two helium collide? They form an isotope of beryllium, which is very unstable. It lasts just barely long enough that there's reasonable chance that a third helium will come along, collide with it, and make carbon. If the lifetime of that beryllium were only a little bit shorter, the Universe would never have gotten beyond helium, and we would live in a Universe of hydrogen/helium, but we wouldn't live there, because the complexity of intelligence would not have been able to form. What a wonderful fluke.

Now, in one of these massive stars, as the star exhausts first its hydrogen, then its helium, and so on, it then begins fusion of more and more heavy elements, and by the processes of nuclear physics, we build up, preferentially, iron in the center, silicon, oxygen, carbon, helium, hydrogen, and these are related somewhat to that curve of binding energy I showed in Lecture Thirty-Four.

Eventually, the massive star exhausts its fuel. A sudden collapse that occurs on a very short timescale happens, and the star rebounds from that collapse and explodes, and that's called a *supernova*. Sometimes, there's a *black hole*, or a neutron star, left after it. That's interesting for astrophysics; but for us, the important point is what the star has done. It's taken all of these heavy elements—the carbon, oxygen, the stuff we are made of—and it has recycled it out into the interstellar medium, where hundreds of millions, billions of years later, it finds itself in these interstellar clouds that are the stellar nurseries, and that stuff begins to condense.

As time goes on, each star generation is cooking up more and more of these heavier elements. It enriches the Universe in these heavier elements, which are needed for things like life. The stars, then, are ultimately what have made the material of which we are made, and that is our most intimate connection to this whole process.

Now, when new stars condense, we now know, and have known since 1995, a process that happened around our star, namely, the formation of smaller, cooler bodies that don't ignite nuclear fusion, and aren't stars, but are planets, as it happens. In 1995, the first extrasolar planet was discovered, and now, we know dozens and dozens of planetary systems. Remarkably, they don't look very much like ours. That's partly a bias, because we tend only to be able to see large planets at this point, or see large planets, Jupiter-sized planets, easily, but we are getting in some surprises there.

If you think about stars like the sun, or any star that isn't giving off lots of X-rays or other harmful radiation, there is some kind of zone called the *habitable zone*, some distance from the star, where conditions for life, at least as we know it—and I have to say that, because we may not know enough about all the ways the Universe can make life—where life can exist. In our own solar system, Venus, Earth, and Mars are all within the habitable zone; but Venus, because of its runaway greenhouse effect, as I discussed in an earlier lecture, cannot support life. Mars, at present, is probably too dry and cold the support life, although as ongoing research is suggesting, it might have once done so.

Let's focus now on Earth. Sometime around the first one billion years, life arose. Pretty soon, we had blue-green algae, and they made the oxygen atmosphere. By one billion years, we had multiple cellular organisms; 400 million years ago, life colonized land. Sixty-five million years ago, the dinosaurs were wiped out with the impact of an asteroid with the Earth, and about four million years ago, the first human beings evolved and developed consciousness.

We are at the end—not the end forever, but the end right now—of a cosmic timeline that looks something like this. Early, we knew nothing. There was a period of cosmic inflation, rapid expansion of the Universe; we had protons and neutrons forming. Helium nuclei formed about half a million years out or a little less, as the origin of the cosmic microwave background. We get galaxies forming at 100 or so million years out. The Earth formed, and the expansion happened to accelerate, about nine billion years out. And, here we are, intelligence evolved, at least on this planet, and who knows how many other places in the Universe.

What about the future? Well, a physicist named Freeman Dyson has written an optimistic paper called, "Time Without End: Physics and Biology in an Expanding Universe," and Dyson argues that once intelligent life get established, the organization that is intelligence will persist to the infinite future, a kind of optimistic assessment.

The recent discoveries—particularly the accelerated expansion—kind of temper Dyson's optimism, but there's still a grand vision of our place as intelligent beings in an ongoing Universe.

Well, this course has been about physics, and how physics is in your life. The most dramatic way physics is in your life is in making the conditions that made your intelligence, your consciousness, and your ability to contemplate the Universe possible.

I want to end by bringing in literature, because all that we do as human beings is linked. We don't just do science, and I think our appreciation of the Universe is enhanced by our knowledge of our connection to it scientifically. I want to end with a favorite poem of mine. I'll just read a sampling from this poem. It's called, "The Origin of Order" by the poet Pattiann Rogers. It's from a wonderful book called, *Verse and Universe*, which is a whole book of poems about science and mathematics. I recommend a number of the poems in there, but I love this one because it evokes this whole sense of cosmic evolution that I've just described in this lecture. It evokes our being, literally, children of the stars, are being made of stardust, and everything around us on planet Earth here, and much of the rest of the visible Universe also being made of stardust.

I'll end with this poem, and I hope you enjoyed this look at physics in your life:

> Stellar dust has settled. It is green underwater now, in the leaves of the yellow crowfoot. Its potentialities are gathered together under the pine litter, as emerging flower of the pink arbutus. It has gained the power to make itself again in the bone-filled egg of osprey and teal. At this moment, there are dead stars seeing themselves as marsh and forest, in the eyes of muskrat and shrew, disintegrated suns making songs all night long in the throats of crawfish frogs.
>
> Child of the sky, ancestor of the sky, the mind has been obligated from the beginning to create an ordered Universe as the only possible proof of its own inheritance.

Thank you, and I hope you enjoyed *Physics in Your Life*.

Glossary

aerodynamic lift: The upward force of air on an airplane or bird wing.

ampere: The unit of electric current, equal to 1 coulomb of charge per second.

amplifier: An electronic circuit that boosts either the voltage or current of an electrical signal.

amplitude: The size of the disturbance that constitutes a wave.

AND: The logical operation whose output is 1 only if both inputs are 1.

angular momentum: A measure of an object's rotational motion; the product of rotational inertia and angular velocity.

angular velocity: A measure of the rotation rate of a rotating object.

antenna: A system of electrical conductors used to send or receive electromagnetic waves.

apparent weight: The "weight" read by a spring scale, which may or may not be your actual weight (the force that gravity exerts on you), depending on whether or not you're accelerating.

apparent weightlessness: The condition encountered in any freely falling reference frame, such as an orbiting spacecraft, in which all objects have the same acceleration and, thus, seem weightless relative to their local environment.

arteriosclerosis: A buildup of fatty plaque in the walls of arteries. Can lead to blockage or to collapse, as described by Bernoulli's principle.

atomic number: The total number of protons in an atom's nucleus and, hence, the number of electrons in a neutral atom. Determines what element an atom belongs to.

axons: Long extensions of neurons that carry signals to other neurons.

battery: A device that converts chemical energy to electrical energy by separating positive and negative charge.

beats: Sound heard at the frequency difference between two sound waves of very similar but not identical frequency.

Bernoulli's principle: A statement of energy conservation in a fluid, showing that the pressure is lowest where the flow speed is greatest and vice versa.

Big Bang: The explosive event that began the Universe as we know it.

bit: A single binary digit, which can have only one of the two values 0 or 1.

buoyancy force: The upward force on an object that is less dense than the surrounding fluid, resulting from greater pressure at the bottom of the object.

byte: A sequence of 8 bits.

cache: Special high-speed computer memory used for temporary storage of data and instructions.

carnot engine: A simple engine that extracts energy from a hot medium and produces useful work. Its efficiency, which is less than 100 percent, is the highest possible for any heat engine.

CCD: See **charge-coupled device**.

center of mass: A point where an object acts as though all its mass were concentrated.

central processing unit (CPU): The main electronic circuitry of a computer, which performs fundamental operations on digital data.

centrifugal force: There's no such thing! Banish this word from your vocabulary. See Lecture Nine.

centripetal force: Any real, physical force that acts to keep an object moving in a circular path. Examples include gravity for the Moon and the friction of tires on the road for a car rounding a curve.

charge-coupled device (CCD): A light detector that captures visual information using electrons in individual picture elements (pixels). Used in digital cameras and many other devices.

chip: See **integrated circuit**.

circular orbit: One of many possible paths for an orbiting object; in a circular orbit, the object remains at a fixed distance from the gravitating center and its speed remains constant.

classical physics: The theories and descriptions of physical reality developed before about the year 1900, specifically excluding relativity and quantum physics.

clock: A circuit inside a computer that generates a periodic signal used for synchronizing and timing all computer operations.

cogeneration: The process of generating both usable thermal energy and electrical energy in the same power plant.

collision: An intense interaction between objects that lasts a short time and involves very large forces.

compression: A technique used to reduce the number of bits needed to store digital information.

conduction: Heat transfer by physical contact.

conductor: A material that contains electric charges that are free to move and can, thus, carry electric current.

Conservation-of-energy principle: The principle that energy cannot be created or destroyed, strictly valid in pre-relativity physics.

conserved quantity: A quantity whose value does not change, at least in a given circumstance.

constructive interference: See **interference**.

convection: Heat transfer resulting from fluid motion.

convection oven: An oven that uses forced circulation of hot air to reduce cooking time.

cosmic microwave background: Electromagnetic radiation in the microwave region of the spectrum, which pervades the Universe and represents a "fossil" relic of the time when atoms first formed, about half a million years after the Big Bang.

cosmological constant: A quantity first introduced by Einstein into his equations of general relativity to provide a kind of antigravity effect that would keep the Universe static; later discredited. Recently revived as a possible explanation for the 1998 discovery that the expansion of the Universe is accelerating.

coulomb: The unit of electric charge.

CPU: See **central processing unit**.

critical mass: The mass of fissile material (uranium, plutonium) needed for a self-sustaining nuclear chain reaction.

curve of binding energy: A graph describing the energy release possible in forming atomic nuclei; shows that both fusion of light nuclei and fission of heavy nuclei can release energy.

data bus: Channel for high-speed data transfer among different components of a computer.

depletion region: The region surrounding a PN junction, in which there is a dearth of free charges.

destructive interference: See **interference**.

differential GPS: Use of two Global Positioning System receivers to reduce timing and atmospheric errors.

diffraction: The phenomenon whereby waves change direction as they go around objects.

diffraction limit: A fundamental limitation posed by the wave nature of light, whereby it is impossible to image an object whose size is smaller than the wavelength of the light being used to observe it.

diffuse reflection: The reflection of waves, especially light, from a rough surface. The light is scattered at different angles and does not form an image.

diffusion: The process where a material or type of particle moves from regions of higher concentration to regions of lower concentration.

digital information storage: The encoding and storage of information as a sequence of digital 0s and 1s.

diode: An electronic device using a PN junction to restrict the flow of electric current to one direction only.

doped semiconductor: A semiconductor to which impurities have been added to alter the material's electrical conductivity.

Doppler effect: The increase in perceived frequency (higher pitch for sound, bluer color for light) of waves when the source approaches the observer. Also, the decrease in frequency when the source recedes from the observer.

drag: The backward-pointing aerodynamic force that resists the forward motion of an airplane, bird, or other heavier-than-air flying object.

dynamic memory: Memory that stores information as electric charge. Must be refreshed several thousand times per second.

elastic collision: A collision in which energy is conserved.

electric charge: A fundamental property of matter that determines electric and magnetic interactions.

electric current: A net flow of electric charge.

electric field: The influence that surrounds an electric charge, resulting in forces on other charges.

electric generator: A device that uses electromagnetic induction to convert mechanical energy to electrical energy. Typically, a generator involves a coil of wire rotating in a magnetic field.

electromagnet: A magnet made by passing electric current through a coil of wire.

electromagnetic induction; A fundamental phenomenon wherein a changing magnetic field produces an electric field.

electromagnetic spectrum: The range of electromagnetic waves, organized by frequency or wavelength.

electromagnetic wave: A structure consisting of electric and magnetic fields, each produced from the change in the other, that propagates through space carrying energy. Light is an electromagnetic wave. In vacuum, all electromagnetic waves travel at exactly the speed of light.

electromagnetism: The branch of physics dealing with electricity and magnetism, described by Maxwell's equations as developed in the mid-19th century.

electromechanical relay: A device using an electromagnetically actuated switch to allow one electric circuit to control another.

electrostatic precipitator: A device that uses electric fields to remove particulate matter from smokestacks.

energy: One of the two basic "things" that makes up the Universe. Energy is what makes everything happen.

energy gap: The range of unavailable energies that separates two bands of allowed energy levels in a semiconductor.

entropy: A measure of disorder. The second law of thermodynamics states that the entropy of a closed system can never decrease.

equatorial orbit: An orbit that remains above Earth's equator.

exclusive OR: The logical operation whose output is 1 if either, but not both, of its inputs is 1.

extrinsic semiconductor: See **doped semiconductor**.

Faraday's law: The mathematical statement describing electromagnetic induction.

FET: See **field-effect transistor**.

field-effect transistor (FET): A transistor in which an electric field exercises the control function.

first law of thermodynamics: The statement that energy is conserved, expanded to include thermal energy.

flip-flop: An electronic circuit that has only two possible states. Used as the fundamental unit in static semiconductor memory.

fluid friction: A friction-like force that slows the flow of a fluid, especially near a solid boundary.

free fall: The state of motion of an object on which the only force acting is gravity. The object need not be moving downward!

frequency: The number of complete wave cycles per unit of time; inverse of the wave period.

friction: A force that acts between two surfaces, opposing any relative motion between them.

fuel cell: A device that combines two chemicals (typically, hydrogen and oxygen), producing electric current in the process.

fusion: A nuclear reaction in which light nuclei join to produce a heavier nucleus, releasing energy in the process.

gate: The controlling electrode of a field-effect transistor; an unrelated definition is a circuit that performs a basic logic function.

general relativity: Einstein's 1915 theory that describes gravity as the curvature of spacetime.

geosynchronous orbit: An equatorial orbit at an altitude of about 22,000 miles, where the orbital period is 24 hours. A satellite in such an orbit remains fixed over a point on the equator.

gigabyte: A measure of computer memory, equal to about a billion bytes (exact value 2^{30}, or 1,073,741,824 bytes).

gravitational lensing: The bending of light by the gravity of massive astrophysical objects.

gravity: A universal attractive force that acts between all objects in the Universe.

greenhouse effect: The trapping of outgoing infrared radiation by certain atmospheric gases, resulting in the warming of a planet.

greenhouse gas: A gas that absorbs infrared radiation, thus contributing to the greenhouse effect.

ground-fault interrupter: A safety device that senses imbalance in current on two wires, then shuts off the circuit to prevent electric shock.

gyroscope: A rapidly spinning object whose rotation axis tends to maintain a fixed orientation.

habitable zone: The region around a star where conditions are appropriate for life as we know it.

half-life: The time it takes for half of the atoms in a sample of radioactive material to decay.

heat capacity: A measure of the energy required to change an object's temperature.

heat pump: A refrigerator run in reverse, pumping heat from the cooler outdoor environment into a building.

hole: A place in a semiconductor where an electron is missing from the crystal structure. Acts as a positive charge.

holographic image: A three-dimensional image made by recording interference patterns of wave fronts coming from the object being imaged.

hyperfine transition: A transition between two very closely spaced atomic energy levels.

induced electric field: An electric field produced not by electric charge but by a changing magnetic field.

insulator: A material with no or few free electric charges and, thus, a poor carrier of electric current.

integrated circuit: A circuit built on a single piece of silicon.

interference: The process whereby two waves, occupying the same place at the same time, simply add to produce a composite disturbance. Interference may be constructive, in which the two waves reinforce to produce an enhanced composite wave, or destructive, in which case the composite wave is diminished.

internal energy: The energy associated with random molecular motion; commonly but mistakenly called "heat."

intrinsic semiconductor: A semiconductor made from a pure material.

ion: An atom that has lost or gained an electron, thus possessing an electric charge.

ionosphere: A region of Earth's atmosphere, beginning about 50 miles up, that contains free electrons and is, therefore, electrically conductive; affects the timing of GPS signals.

kinetic energy: The energy associated with an object's motion.

lagrangian point, L1: A point roughly 1 million miles sunward of Earth, where a spacecraft's orbital period is 1 year, allowing it to stay on the line between Earth and Sun.

laser: A device that produces light or other electromagnetic radiation through stimulated emission; stands for *L*ight *A*mplification by *S*timulated *E*mission of *R*adiation.

laser angioplasty: The use of laser beams to clear clogged arteries by vaporizing plaque.

latent heat: Energy associated with a substance's being in a state requiring higher energy, as in the latent heat of water vapor, which can be released when the water condenses.

lLaw of inertia: The statement that a body in motion (or at rest) remains in uniform motion (or at rest) unless a force acts on it.

LCD: See **liquid-crystal display**.

LED: See **light-emitting diode**.

lens: A piece of transparent material shaped so that refraction brings light rays to a focus.

lift: See **aerodynamic lift**.

light-emitting diode (LED): A diode engineered to produce visible or near-visible light when current flows across its PN junction.

liquid-crystal display (LCD): A visual display device that uses electric fields to reorient the molecules of a liquid crystal, thereby altering the polarization of light.

low-Earth orbit: An orbit whose altitude above Earth's surface is a small fraction of Earth's radius. The period of low-Earth orbits is about 90 minutes.

magnetic field: The influence surrounding a moving electric charge (and, thus, a magnet) that results in forces on other moving charges (and on magnets or magnetic materials).

magnetic resonance imaging (MRI): A procedure that uses spinning protons in a magnetic field to form images of the body's interior.

magnetron: A special vacuum tube in which electrons undergo circular motion, producing microwaves.

Maxwell's equations: A set of four equations that describe all electromagnetic phenomena of classical physics.

mechanics: The study of motion.

megabyte: A measure of computer memory, equal to about a million bytes (exact value 2^{20}, or 1,048,576 bytes).

memory: An electronic circuit that maintains a given state until the state is explicitly changed.

metal-oxide-semiconductor field-effect transistor (MOSFET): A type of transistor widely used in computer circuits.

microprocessor: The single-chip CPU of personal and other small computers.

minority charge carriers: The free charges that are in a minority in a given semiconductor (electrons in P type, holes in N type).

mirage: An image formed by refraction because of temperature gradients in the air.

moderator: In a nuclear reactor, a substance that slows neutrons to make them more effective at causing fission.

modern physics: The theories and descriptions of physical reality developed after about the year 1900, including specifically, relativity and quantum physics.

momentum: A quantity that describes the "amount of motion" in a moving object, accounting for both velocity and mass.

Moore's law: The statement that the number of transistors per integrated circuit grows exponentially, doubling every year or two. Moore's law has held since the 1960s.

MOSFET: See **metal-oxide-semiconductor field-effect transistor**.

motherboard: A circuit board holding the CPU, memory, and other components central to the operation of a computer.

MRI: See **magnetic resonance imaging**.

NAND: NOT AND; the logical operation whose output is the opposite of AND.

natural greenhouse effect: The effect of natural greenhouse gases, particularly water vapor and carbon dioxide, in raising Earth's temperature some 60° F above what it would otherwise be.

negative charge: The type of electric charge on the electron.

net force: The sum of all forces acting on an object.

neurons: Specialized cells that transmit electrochemical signals in the brain and nervous system.

neutral buoyancy: The state of neither rising nor sinking that occurs for an object of the same density as the surrounding fluid.

neutron: An electrically neutral component of the atomic nucleus.

neutron activation: A process of inducing artificial radioactivity by bombarding substances with neutrons; the subsequent radioactive decay is used to identify the substances.

Newton's first law of motion: This is the same as the law of inertia.

Newton's second law of motion: The statement that an object's acceleration is proportional to the net force applied to it and inversely proportional to its mass.

Newton's third law of motion: The statement that forces always come in pairs; if one object exerts a force on a second object, the second exerts an equal but opposite force back on the first.

nonthermal energy transfer: Energy transfer that does not rely on a temperature difference, as in a microwave oven.

nonvolatile memory: Memory that retains information even when the power is off, as in a digital camera.

NOR: NOT OR; the logical operation whose output is the opposite of OR.

NOT: The logical operation whose output is the opposite of its input.

N-type semiconductor: A semiconductor doped so that the dominant free charges are negative electrons.

nuclear chain reaction: An ongoing reaction in which neutrons released in nuclear fission go on to cause additional fission events.

nuclear force: The force that binds protons and neutrons to form atomic nuclei.

nuclear magnetic resonance (NMR): The process at the heart of MRI, whereby protons absorb radio waves of just the right frequency to set them precessing in a magnetic field.

nuclear medicine: The use of radioactive substances to image body structures and analyze physiological processes.

nucleosynthesis: The process of forming atomic nuclei, especially in stars and in the early Universe.

Ohm's law: The statement, valid for some materials, that the electric current is proportional to the applied voltage and inversely proportional to the material's resistance.

optical storage medium: A medium, such as the CD or DVD, that encodes information in ways that can be read using light.

optics: The branch of physics dealing with light and its behavior.

OR: The logical operation whose output is 1 if either or both inputs are 1.

pacemaker: A specialized group of cells that provides the signal to govern the rhythmic beating of the heart.

parallel communications: Data transfer that moves many bits simultaneously on separate wires.

period: The time interval between two successive wave crests; equivalently, the time for a complete wave cycle.

PET: See **positron emission tomography**.

phase change: A change in a material, as from solid to liquid or liquid to gas, that occurs abruptly at certain values of temperature and pressure.

phase diagram: A diagram showing how the phases of a substance relate to its temperature and pressure.

photolithography: A process using light to lay down patterns for forming integrated circuits.

photovoltaic cell: A semiconductor device that converts light directly into electrical energy.

piezoelectric device: A device using a material that generates electricity when squeezed or distorted; conversely, the device changes size or shape when a voltage is applied to it.

pixel: An individual element of a digital image.

plasma: An ionized gas, sometimes called the "fourth state of matter."

PN junction: A junction of P- and N-type semiconductors, with the property that electric current can flow in only one direction.

polar orbit: An orbit that passes over Earth's poles. As Earth rotates, a satellite in polar orbit passes over every point on the planet.

polarization: The direction of an electromagnetic wave's electric field.

population inversion: A situation in which more higher level atomic states are populated than are lower level states. Needed for laser action.

positive charge: The type of electric charge on the proton.

positron emission tomography (PET): A medical imaging technique using gamma rays from the annihilation of positrons (anti-electrons) released in the decay of radioactive substances.

potential energy: Stored energy associated with a configuration of objects.

power: The rate of producing or expending energy. In electrical devices, power is the product of voltage and current.

precession: The gradual change in direction of a rotating object's rotation axis as a result of an applied torque.

proton: A positively charged component of the atomic nucleus.

P-type semiconductor: A semiconductor doped so that the dominant free charges are positive holes.

pulsar: A rapidly spinning neutron star.

quantum computing: Computing based on the states of quantum-mechanical systems.

quantum physics: The theory, developed in the early 20th century, that describes physical reality at the atomic scale and below. In this realm, the discrete, "quantized" nature of both matter and energy become important.

radiation: Heat transfer by electromagnetic waves.

RAM: See **random-access memory**.

random-access memory (RAM): Memory whose individual storage locations can all be accessed in equal time, as opposed to sequential memory, such as that on magnetic tape.

read-only memory (ROM): Memory whose state cannot be changed.

rechargeable battery: A battery in which the passage of electric current from an outside source results in the storage of chemical energy.

reflection: The phenomenon whereby a wave strikes a material and rebounds at the same angle with which it struck the material.

refraction: The phenomenon of waves changing direction of propagation when going from one medium to another.

resistance: The property of a material that describes how it impedes the flow of electric current.

resistor: A device formulated to have a specific electrical resistance.

reverse bias: The condition in which a voltage is applied across a PN junction, with positive to the N-type side. Results in very little electric current.

ROM: See **read-only memory**.

rotational inertia: A measure of an object's resistance to change in rotational motion.

second law of thermodynamics: A general principle stating that systems tend to evolve from more ordered to less ordered states.

semiconductor: A material that lies between insulators and conductors in its capacity to carry electric current. The electrical properties of semiconductors are readily manipulated to make the myriad devices at the heart of modern electronics.

semiconductor memory: Memory made with transistors and other devices. The fastest memory used in computers.

serial communications: Data transfer that moves one bit at a time, using a single wire.

shock wave: A very strong, abrupt wave produced when a wave source moves through a medium at a speed faster than the waves in that medium. An example is a sonic boom from a supersonic airplane.

sliding friction: The frictional force between two surfaces in relative motion; smaller than static friction.

special relativity: Einstein's 1905 theory that shows how all uniformly moving frames of reference are equivalent as far as the laws of physics are concerned. Requires modification of our commonsense notions of time and space.

specular reflection: Reflection off a smooth surface that appears shiny and produces an image, as in a mirror.

spontaneous emission: The emission of light or other electromagnetic energy as an electron jumps spontaneously from a higher energy level to a lower one.

standing waves: Waves that "stand" without propagating on a medium of fixed size. The vibrations of a violin string are standing waves.

static electricity: Electricity associated with stationary distributions of electric charge.

static friction: The frictional force between two surfaces at rest relative to each other.

static memory (SRAM): Semiconductor memory in which information is stored in the states of flip-flops.

steady-state theory: The idea, now widely discredited, that the overall structure of the Universe never changes.

stimulated emission: The emission of light or other electromagnetic energy as an electron jumps from a higher energy level to a lower one, stimulated to do so by the nearby passage of similar electromagnetic energy.

sublime: To change directly from solid to vapor, without going through the liquid state.

superconductor: A material that, at sufficiently low temperature, exhibits zero resistance to the flow of electric current.

superheated: A liquid above its boiling point but nevertheless not boiling.

supernova: The violent explosion marking the endpoint of massive stars.

temperature: A measure of the average thermal energy.

terminal speed: The maximum speed reached by a falling object, which occurs when air resistance becomes equal in magnitude to the force of gravity.

Theory of Everything: An as-yet-undeveloped theory that would describe all of physical reality.

thermal energy: See **internal energy**.

thermal energy balance: A state wherein energy leaving a system is balanced by incoming energy.

thermal pollution: Waste heat dumped to the environment, usually associated with the thermodynamic inefficiency of power plants.

thermistor: A temperature-measuring device utilizing the property that the resistance of an intrinsic semiconductor decreases with increasing temperature.

thermocouple: A device that uses the thermoelectric effect to measure temperature.

thermodynamics: The branch of physics dealing with heat and related phenomena.

thermoelectric effect: The production of a voltage at a junction of two dissimilar materials when heated.

toner: The small particles that take the place of ink in dry copying and laser printing (xerography).

torque: The rotational analog of force; torque depends on force and where that force is applied.

total internal reflection: Complete reflection that occurs as light attempts to go from a more dense to a less dense medium, as from water to air.

transformer: A device that uses electromagnetic induction to transform high-voltage/low-current electricity to low-voltage/high-current and vice versa.

transistor: A semiconductor device with three separate electrical connections, in which current or voltage in one circuit controls current or voltage in another circuit. The basic control element in both digital and analog electronics.

truth table: A table that displays all possible states of a logic gate.

volatile memory: Memory that stores information only as long as power is applied.

voltage: A measure of the energy per unit of electric charge.

watt: A unit of power, equal to 1 joule of energy per second.

wave: A traveling disturbance that carries energy but not matter.

wavelength: The distance between two successive wave crests.

weight: The force that gravity exerts on an object.

word: A sequence of binary bits, usually 32 or 64 bits, on which a computer performs operations.

working fluid: A substance used in refrigerators and engines to transfer heat; often undergoes phase changes in the process.

XOR: See **exclusive OR**.

Bibliography

Armenti, Angelo, ed. *Physics of Sports*. New York: American Institute of Physics, 1992. This book is a compendium of articles, many from *The American Journal of Physics*, on the physical principles behind individual sports. The author has added introductory essays on each topic.

Barham, Peter. *The Science of Cooking*. New York: Springer, 2001. A fascinating look at scientific principles at work in the kitchen. Separate chapters cover such topics as fish, breads, sauces, sponge cakes, and many more. Complete with recipes.

Bloomfield, Louis A. *How Things Work: The Physics of Everyday Life*. New York: Wiley, 1997. A nice blend of general physics principles and specific applications. The main chapters have physics titles, such as "Simple Mechanical Objects" and "Thermodynamics," but subheadings cover particular technologies, including "Clothing and Insulation," "Water Faucets," or "Seesaws." A great reference source for understanding our technological world.

Brain, Marshall. *Marshall Brain's How Stuff Works*. New York: Hungry Minds, Inc., 2001. A handsomely illustrated "how things work" book treating such catchy general categories as "In the Air," "On the Road," "Of Microprocessors, Mice, and Modems," "Picture Perfect," "Around the House," and so forth. Within the categories, individual technologies seem to have been chosen somewhat at random, but what's covered is generally done well. See also the web site www.howstuffworks.com and the 2002 sequel *More How Stuff Works*.

Chaisson, Eric. *Cosmic Evolution: The Rise of Complexity in Nature*. Cambridge, MA: Harvard University Press, 2001. A cross-disciplinary look at the evolution of the Universe based on the underlying theme of ever-increasing complexity. Chaisson even develops a mathematical measure of complexity that he applies to his cosmic history.

Dalton, Stephen. *The Miracle of Flight*. Buffalo, NY: Firefly Books, 1999. Lovely photos of all manner of birds, insects, and aircraft in flight make for an artful book, while plenty of scientifically accurate drawings round out the presentation.

Davidovits, Paul. *Physics in Biology and Medicine*, 2nd edition. San Diego: Harcourt Academic Press, 2001. A survey of physics principles applied mostly to human physiology and medical technology. Does not shy away from math but limits the math to simple algebra.

El-Rabbany, Ahmed. *Introduction to GPS: The Global Positioning System.* Boston: Artech House, 2002. A comprehensive and eminently readable introduction to the workings of the Global Positioning System.

Enge, Per. "Retooling the Global Positioning System," *Scientific American,* May 2004, pp. 91–97. The Global Positioning System is continually refining its capabilities, through improvements on the ground and in space. This article shows what GPS users can expect as GPS evolves in the coming years.

Ferendeci, Altan. *Physical Foundations of Solid State and Electron Devices.* New York: McGraw-Hill, 1999. Heavy slogging through the physics and math of semiconductor devices. If you really want the details, here they are.

Fountain, Henry. *The New York Times Circuits: How Electronic Things Work.* New York: St. Martin's Press, 2001. Another "how things work" book with emphasis on electronics. Same level and general style as *Marshall Brain's How Stuff Works.*

Garwin, Richard, and Georges Charpak. *Megawatts and Megatons: A Turning Point in the Nuclear Age?* New York: Knopf, 2001. Two experts on nuclear issues speak with authority on nuclear power and weapons. Charpak is a French Nobel laureate in physics, while Garwin won the prestigious Enrico Fermi Award.

Griffith, W. Thomas. *The Physics of Everyday Phenomena: A Conceptual Introduction to Physics,* 4th edition. Boston: McGraw-Hill, 2004. This book is more slanted toward basic physics than its title might imply. It's basically an introductory-level college physics textbook using some algebra but no calculus. Plenty of real-world applications are included.

Grubbs, Bruce. *Using GPS: GPS Simplified for Outdoor Adventurers.* Helena, MT: Falcon, 1999. This practical handbook by an avid hiker, biker, and cross-country skier provides a practical complement to the El-Rabbany book's technical exposition of the Global Positioning System.

Guth, Alan. *The Inflationary Universe.* Reading, MA: Addison-Wesley, 1997. Guth, one of the two inventors of the inflationary universe concept, presents a readable history of modern cosmology with emphasis on his involvement and the discovery of inflation. Although recent, the book is missing significant developments in cosmology that began in 1998 with the discovery of accelerated expansion.

Hewitt, Paul. *Conceptual Physics*, 9th edition. San Francisco: Addison Wesley, 2002. This popular textbook for nonscience physics courses emphasizes a conceptual, as opposed to quantitative, understanding of physics.

Horowitz, Paul, and Winfield Hill. *The Art of Electronics*, 2nd edition. New York: Cambridge University Press, 1989. The "bible" for anyone seeking to understand or design electronic circuits.

Kaler, James B. *Cosmic Clouds: Birth, Death, and Recycling in the Galaxy*. New York: Scientific American Library, distributed by W.H. Freeman and Co., 1997. A detailed look at the stellar recycling process that connects us and the materials we're made from to the broader cosmos. Illustrated with magnificent color photos, many from the Hubble Space Telescope.

Kevles, Bettyann. *Naked to the Bone: Medical Imaging in the Twentieth Century*. Reading, MA: Addison Wesley, 1998. A history of medical imaging from the earliest x-rays to magnetic resonance and other modern techniques. Includes a multipage timeline.

Krauss, Lawrence. *Atom: An Odyssey from the Big Bang to Life on Earth...and Beyond*. Boston: Little, Brown, 2001. Physicist and popular science writer Krauss tells the tale of cosmic evolution from an atom's viewpoint.

Laws, Kenneth. *Physics and the Art of Dance*. New York: Oxford University Press, 2002. Laws, a physicist who learned classical ballet, describes the physical principles behind common dance movements. Nicely illustrated with photos and diagrams.

Maiman, Theodore. *The Laser Odyssey*. Blaine, WA: Laser Press, 2000. The history of the laser is fraught with controversy. Theodore Maiman, author of this book, is commonly cited as having developed, built, and patented the world's first practical laser. This is his autobiographical story of the laser's history—but see also the book by Townes.

McGraw-Hill Encyclopedia of Science and Technology. New York: McGraw-Hill, updated yearly. A deep and authoritative resource for understanding scientific phenomena and their technological applications.

Pohlmann, Ken C. *The Compact Disc: A Handbook of Theory and Use*. Madison, WI: A.R. Editions, Inc., 1989. A thorough but nevertheless nicely readable introduction to the workings of CD technology.

Schneider, Stephen H., Armin Rosencranz, and John O. Niles, eds. *Climate Change Policy: A Survey*. Washington, DC: Island Press, 2002. The first chapter provides a concise introduction to climate science and global warming.

Scientific American. New York, Scientific American, Inc., monthly. This popular yet authoritative magazine features frequent articles on science and its technological applications. The monthly feature "Working Knowledge" describes a particular technology or device in detail.

Scientific American, fall 1997, special issue, "The Solid-State Century." New York: Scientific American, Inc. This special issue features articles on a wide range of electronic technologies and devices.

Sky and Telescope, October 2003, special issue, "The New Cosmology." Cambridge, MA: Sky Publishing. This issue of the popular astronomy magazine brings the reader up to date on the stunning cosmological discoveries of the late 1990s and early 2000s.

Snow, C. P. *The Two Cultures and the Scientific Revolution*. New York: Cambridge University Press, 1959. This attempt to bridge the cultural gap between the sciences and the humanities provides the title quote for this course's lecture on the second law of thermodynamics.

Sochurek, Howard. *Medicine's New Vision*. Easton, PA: Mack Publishing, 1988. Noteworthy for its lavish full-page color images.

Storey, Neil. *Electronics: A Systems Approach*, 2nd edition. Harlow, England: Addison Wesley Longman, 1998. This textbook offers a good introduction to electronic devices and circuits, including the logic circuits discussed in Module Four. Some math included.

Taylor, Jim. *DVD Demystified*. New York: McGraw-Hill, 2000. History, features, and technology behind the digital versatile disc.

Townes, Charles H. *How the Laser Happened: Adventures of a Scientist*. New York: Oxford University Press, 1999. In the 1950s, Charles Townes and his colleague Arthur Schawlow invented a microwave version of the laser (the maser). They later speculated on the possibility of a similar device producing optical emission. Although Townes and Schawlow received a patent and won the Nobel Prize, there remains controversy about who really originated the idea and built the first laser. This book is Townes's autobiographical memoir emphasizing the history of the maser/laser.

Vogel, Steven. *Life's Devices: The Physical World of Animals and Plants*. Princeton, NJ: Princeton University Press, 1988. A nice introduction to physical principles in biology.

Wegener, Peter. *What Makes Airplanes Fly?* New York: Springer-Verlag, 1991. A historical and scientific account of human flight. Occasional math included.

Wolfson, Richard. *Nuclear Choices: A Citizen's Guide to Nuclear Technology.* Cambridge, MA: MIT Press, 1993. Written for nonscientists, this book presents the technology of nuclear power and nuclear weapons in an unbiased way that lets the reader form his or her own conclusions.

Wolfson, Richard, and Jay M. Pasachoff. *Physics for Scientists and Engineers*, 3rd edition, extended version with modern physics. Menlo Park, CA: Addison Wesley, 1999. A standard introductory physics text for students of science and engineering. Useful for a deeper understanding of the basic physics behind the topics discussed in this course. Uses calculus.

Wolke, Robert L. *What Einstein Told His Cook: Kitchen Science Explained.* New York: W.W. Norton, 2002. An immensely entertaining account of kitchen science, complete with recipes.

Wood, Elizabeth. *Science from Your Airplane Window*. New York: Dover Publications, 1975. Quick explanations of many phenomena associated with airplane flight; especially strong on the optical phenomena you see from the plane's window. Take this book on your next plane trip!

Internet Resources:

www.howstuffworks.com. This is the web site associated with Marshall Brain's book of the same title; see above. Here, you can find illustrated descriptions of the workings of almost any technological device or natural phenomenon.

www.nsdl.org. This is the National Science Digital Library, sponsored by the National Science Foundation. The site has a search engine that turns up links to sites on scientific and technological topics.

www.sciam.com/askexpert_directory.cfm. *Scientific American*'s "Ask the Experts" page lets you pose questions to scientific and engineering experts, or you can read answers to others' questions.

Notes

Notes

Notes